普通高等教育“十四五”规划教材

Sustainable Chemical Engineering towards Zero-Carbon Goals

面向碳中和的“绿色化工＋”

金　鑫　宁　汇◎主　编
严文娟　曹　宁　祝晓琳◎副主编
Raghunath V.Chaudhari ◎校　核

中国石化出版社
·北京·

图书在版编目（CIP）数据

面向碳中和的“绿色化工+”= Sustainable Chemical Engineering towards Zero-Carbon Goals：英文 / 金鑫，宁汇主编 . -- 北京：中国石化出版社，2024. 9.
(普通高等教育“十四五”规划教材) . -- ISBN 978-7-5114-7533-6

Ⅰ. TQ02

中国国家版本馆 CIP 数据核字第 20247BV166 号

中国石化出版社出版发行

地址：北京市东城区安定门外大街 58 号
邮编：100011　电话：(010)57512500
发行部电话：(010)57512575
http://www.sinopec-press.com
E-mail：press@sinopec.com
宝蕾元仁浩（天津）印刷有限公司印刷
全国各地新华书店经销

*

710 毫米 ×1000 毫米 16 开本 20.75 印张 357 千字
2024 年 9 月第 1 版　2024 年 9 月第 1 次印刷
定价：79.00 元

Dedicated to my mentors

who have been teaching me to work on problems that matter.

前 言

化学工业是国民经济发展的基石，从能源生产到基础化学品制造，从环境保护到航空航天，从固碳降耗到医药健康，化学工业在与人民休戚相关的各个领域扮演着至关重要的角色。培养具有国际视野、胜任未来化学工业技术挑战的后备化工人才，是化工行业向高端化、智能化和绿色化转型的关键所在。当前化工行业已经呈现出前所未有之剧变，从重视化学、数学和物理等基本理论的应用，到以“化工+”为特色的学科交叉融合创新，未来化工行业的发展急需培养兼具跨领域知识技能的创新型人才。培养能够开展“文理结合、理工融合和工工交叉”的复合型高端人才，是未来化工类专业人才培养的重要目标。然而，当前化工类专业大学生培养过程仍面临“学生兴趣不足、知识结构碎片化、学科交叉跨而不深”等一系列难题。如何在基础课程学习阶段开展切实有效的专业交叉通识教育，如何在专业课学习阶段提升学科交叉学习和创新能力，是摆在化工教育者面前的重要课题。

笔者根据在国内外多年的科研与育人实践经验，从国外优秀通识教育和专业教材中选取了部分精华内容，编写了面向化工类专业低年级大学生的新形态教材。本书兼具“专业导论”、专业通识教育、学术英语和跨学科前沿等多重功能，以相关行业历史发展、基础知识与概念和科学技术革新为基本教学脉络，激发大一新生从事化工类交叉学科学习的兴趣，在入学初就建立“大化工”观；进而引导大二学生构筑“化工+”的通识知识体系，在开始学习专业课之前，培养能够进行“化学–材料–过程”多学科交叉研学的能力。尽管采用英语作为基本教学语言，但是本书中行文与基本用词可以满足大学生大学外语与学术外语衔接阶段从一级到四级的学习需求，在学习化学、物理和数学课之外接触并了解交叉领域的重要概念，为从事“化工+”交叉领域研究打下坚实的基础。同时本教材可作为没有化工背景但立志从事解决化工行业重大科技难题的研究生，快速了解

“大化工”思维的参考书。本教材首先介绍了化工基本物料衡算与能量衡算，提升大学生“化学–物理–数学”融合培养环节成效；引入工程伦理概念，引导学生在开始专业课学习之前构建健康的伦理观；太阳能、生物质利用、石油加工和电池等章节均涉及化学、材料和工艺过程；生命周期分析章节融合了前几章对“大化工”的认识理解，作为未来化工复合人才培养重要的落脚点。

本教材在编写过程中得到了中国石油大学（华东）杨朝合教授与美国堪萨斯大学Raghunath V. Chaudhari院士的指导与帮助。编写团队本身就具备能源化学、能源材料、能源化工和能源产品等研究背景，在每个章节的选取与编写过程中充分考虑到新时代化工类专业学生的学习方法和模式。

本书中附带多个二维码，相关线上资源会根据行业发展需求定期更新，并利用AI工具帮助学生形成动态知识导图，真正体现了新形态教材“交互并迭代”的显著特色，期待本教材能够成为新时代大学生的入门启蒙书、学习参考资料和创新思想的孵化器。

由于编写者能力所限，如本书有不明晰和欠妥之处，还请各位专家和读者多斧正。

Preface

This textbook has been edited for undergraduate and graduate students with chemical engineering, chemistry, materials and engineering safety background. This book was edited with the aim of providing fundamentals for engineering students, particularly for exploring interdisciplinary topics and solving engineering science problems in both academic and industrial endeavors. The idea of compiling those chapters originated from the fact that, most engineering students are lack of professional training for scientific writing and analytical thinking. Therefore, this textbook will be suitable as a reference reading for junior or senior students in "Professional English", "Scientific Writing" as well as "Introduction to Renewable Energies", as well as in "Introduction to Chemical Engineering".

Completely different from classic introductory textbooks for chemistry related students, this book first discusses the general terminologies and ethics in chemical engineering fields. It is very important, as most existing textbooks focuses on technical training while the "academic spirit" and "human" parts are not "there" in earlier career for most engineering talents. Numerous incidents and academic misconduct reports across the globe have motivated the authors to find a new way to establish the human parts prior to teaching engineering knowledge and skills.

The dramatic evolution in energy distribution and allocation across the world during pandemic time reveals the following truth in engineering education: breaking the border among "chemical engineering + X" has never been so critical and urgent, as pharmaceutical innovation, material processing, energy revolution and equipment upgrading all demand immediate transformation from traditional college education to interdisciplinary training (cultivation). "There is plenty of room at the borders."

Here is the summary for major aspects in this book:

(1) General knowledge for chemical engineering as a subject.

(2) The relationship between engineering and technology, the teaching innovation and academic integrity.

(3) General principles for solar energy transformation to electricity and chemical energy, water splitting for hydrogen and oxygen evolution.

(4) Composition of biomass and conversion to fuels and chemicals.

(5) Composition of crude oils, refining processes and downstream petrochemicals.

(6) Structures and applications of batteries.

(7) Fundamentals of fuel cells, characterization of performances and oxygen reduction reaction.

(8) The concept of life cycle assessment for sustainable technologies and economies.

The editors want to acknowledge the guidance and support from Dr. Raghuanth V. Chaudhari (Distinguished Professor Emeritus at The University of Kansas), Dr. Bala Subramaniam (Distinguished Professor at The University of Kansas), Dr. Chaohe Yang (Distinguished Professor at China University of Petroleum).

My father, Mr. Youhai Jin (Distinguished Professor Emeritus at China University of Petroleum) has provided valuable suggestions not only to this book, but for my career path as a good teacher, mentor, researcher in the field of green engineering. I myself particularly appreciate the support, trust and dedication from the doctorate and master students in Nanocatalysis for Biomass Conversion Research Lab at China University of Petroleum.

Engineering Chemistry for Innovation.

Leadership for good starts here.

Jin Xin（金鑫）

Guzhenkou, Qingdao

Contents

Chapter 1 Principles in Chemical Engineering（化学工程的基本概念和原理）

"Fascinating! So that's why chocolate-chip cookie dough tastes better raw!"

Part A Comprehensive Reading

Introduction to Chemical Engineering Profession

1 Definitions of Chemical Engineering

There is no universally accepted definition of the field of chemical engineering to date. A definition found in a standard dictionary states, “Chemical engineering is a branch of engineering that involves the design and operation of **large-scale** chemical plants, petrochemical refineries, and the like.” Another definition from the information highway goes like this, “Chemical engineering is concerned with processes that cause substances to undergo required changes in their **chemical or physical composition**, structure, energy content, or physical state.” A more appropriate definition of chemical engineering states:

Chemical engineering is the profession in which a knowledge of mathematics, chemistry, biology and other natural sciences gained by study, experience, and practice is applied with judgment to develop economic ways of using materials and energy for the benefit of mankind. The profession encompasses the spectrum from products, to the processes and equipment for making them, and to their applications.

Despite the lack of consensus among chemical engineers on the definition of their **discipline**, which may be attributed to the broad nature of the discipline itself, there is no disagreement that chemical engineers turn low-value materials into high-value products are involved in product design and development; design processes to manufacture products; are involved in process **scale-up**, development, and **optimization**; perform economic analysis of the production process; operate and control the processes to ensure that product quality satisfies the required specification;

and are involved in the management of the processes and in product sales and technical service.

Because of the reliance of chemical engineering on principles and laws that are versatile in the natural world, the possibilities offered to by chemical engineering as well as the challenges faced by chemical engineers are quite formidable and constantly changing. Without intending to be exhaustive, the list of chemical engineering applications includes:

✓ Traditional areas, for example, mining, pulp and paper, oil refining, materials (rubber, plastics, etc.), and environment.

✓ Nontraditional areas such as microelectronics (semiconductor manufacturing), biotechnology (pharmaceutical production processes, genetic engineering, etc.), and nanotechnology.

✓ Other areas, for example, medicine, consulting, law, and business.

A similarity among all chemical engineering systems is that they involve processes designed to **transform** raw materials (generally, low value) into desired products (with added value). A typical problem in the design of a new process or the modification of an existing process is as follows: "Given the amount and the properties of the raw materials, calculate the amount of the products and determine their properties or vice versa." In order to solve this problem, chemical engineers have powerful tools at their disposal that consist of the principles of material and energy balances (and momentum balance, which is not covered in this text), and a broad array of fundamental sciences.

2 Sets of Units and Unit Conversion

Chemical engineers, like many other engineers, use values, units, and dimensions all the time. For example, in a statement like "2.0 L of drinking water…," the "value" (2.0) is the numerical figure, the "unit" (L) indicates what the measured quantity represents, and the "**dimension**" is the measurable quantity that the unit represents. In this case, liter is a unit of volume. A measured or counted quantity has a numerical value (e.g., 2.0) and a unit (e.g., L). It is necessary in engineering calculations to report both the value and the unit. A value without its unit is meaningless. A dimension is a property that can be measured, for example, length, mass, time, and temperature, or calculated by multiplying or dividing other dimensions, for example, length/time=velocity, length^3=volume, and mass/length^3=density. Measurable units are specific values of dimensions that have been defined by convention, such as grams for mass, seconds for time, and centimeters for length. Units can be treated like algebraic variables when quantities are added, subtracted, multiplied, or divided. Derived units can be obtained by multiplying or dividing base units (i.e., units of length, mass, time, and temperature, see Table 1 and 2). There are several systems of units, but two primary systems that engineers use are the International System of Units (SI system) and the American Engineering System of Units (AES). Other systems are the centimeter-gram-second (CGS), foot-pound-second (FPS), and the British System of Units (British). Table 3 shows the units associated with these systems of units.

Table 1. Quantities. Units, and Symbols of the SI System

Quantity	Unit (Base Unit)	Symbol
Length	Meter	m
	Centimeter	cm
Mass	Kilogram	kg
	Gram	g
Time	Second	s
	Day	day
Temperature	Celsius	℃
	Kelvin	K

Source: Reklaitis, G.V., *Introduction to Material and Energy Balances*, John Wiley & Sons, New York. 1983.

(1) Conversion of Units

A measured quantity can be expressed in terms of units having the appropriate dimension. For example, velocity may be expressed in terms of feet per second, miles per hour, centimeters per year, or any other ratio of length and time. The numerical value of velocity depends on the units chosen, for example, 20m/s=66ft/s. To convert a quantity expressed in terms of one unit to its equivalent in terms of another unit, you will need to multiply the given quantity by the conversion factor (new unit/old unit).

Table 2. Units for Mass, Length, and Time

Quantity	Unit (Base Unit)	Symbol	In Terms of Base Units
Volume	Liter	L	0.001 m^3
Force	Newton	N	1（kg·m）$/s^2$
Energy	Joule	J	1N·m=1（kg·m^2）$/s^2$
Pressure	Pascal	Pa	N/m^2
Density	—	ρ	g/cm^3
Molecular weight	—	M	g/mol

Sources: Felder, R.M. and Rousseau, R.W. *Elementary Principles of Chemical Processes*, 3rd den., John Wiley & Sons, New York, 1999; Himmelblau, D.M., *Basic Principles and Calculations in Chemical Engineering*, 3rd edn., Prentice-Hall, Englewood Cliffs, NJ, 1974.

Table 3. Units Associated with Systems of Units

System	Mass (m)	Length (l)	Time (t)	Temperature (T)
SI	Kilogram (kg)	Meter (m)	Second (s)	Kelvin (K)
AES	Pound mass (lb_m)	Foot (ft)	Second (s)	Degree Fahrenheit (℉)
CGS	Gram (g)	Centimeter (cm)	Second (s)	Kelvin (K)
FPS[a]	Pound mass (lb_m)	Foot (ft)	Second (s)	Degree Fahrenheit (℉)
British	Slug	Foot (ft)	Second (s)	Degree Celsius (℃)

[a] Imperial system units are sometimes referred to as FPS.

(2) Temperature Measurement

Temperature is a measure of the average kinetic energy of a substance. We can measure temperature using physical properties of a substance, which change as a function of temperature. Such properties include volume of a fluid (thermometer), resistance of a metal (resistance thermometer), voltage at the junction of two dissimilar metals (thermocouple), and spectra of emitted radiation (pyrometer).

Common temperature scales are listed as follows:

(a) Fahrenheit: On this scale, the freezing point of water is 32 and boiling point is 212.

(b) Celsius: The freezing point of water is 0 and boiling point is 100.

(c) Rankine: On this scale, the absolute zero (when all kinetic energy vanishes) is 0, 459.67 Rankine increment=0 Fahrenheit increment (459.67°R=0℉).

(d) Kelvin: The absolute zero (when all kinetic energy vanishes) is 0, 273.15 Kelvin increment=0 Celsius increment (273.15K=0℃).

3 Significant Figures

The **significant figures** of a number are the digits from the first nonzero digit on the left to either the last digit (zero or nonzero) on the right, if there is a decimal point, or the last nonzero digit of the number, if there is no decimal point. The number of significant figures is readily seen if scientific notation is used. Scientific notation is generally a more convenient way of representing large numbers.

Numbers can be written in either standard notation or scientific notation to show significant figures.

(1) With decimal point: Count from nonzero on the left to the last digit on the right.

0.0012 (1.2×10^{-3}): 2 significant figures

0.001200 (1.200×10^{-3}): 4 significant figures

(2) Without decimal point: Count from nonzero on the left to the last nonzero on the right.

40500 (4.05×10^{4}): 3 significant figures

(3) Scientific notation: Typically only significant figures are shown.

1.47×10^{8}: 3 significant figures

4.0500×10^{4}: 5 significant figures

There is an accepted convention regarding the reporting of computed results. It is simply that a final result should be listed to no more significant figures than the least accurate factor in the expression evaluated. For example, the distance traveled by a vehicle moving at a velocity of 16.223m/s in 2.0s is 32m. The fact that your calculator may list 32.446000m as the result of the computation is an electronic artifact. It is your job to round the result to an appropriate number of significant figures. Intermediate calculations should be performed with the full precision of the tool you are using (calculator or computer), but the result must reflect the true precision. The rules that apply to the result of multiplication/division and addition/subtraction operations are explained next.

4 Moles and Molecular Weight

Atomic weight is the mass of an atom of an element. Mole is the amount of the species whose mass in grams is numerically equal to its molecular weight. One mole of any species contains approximately 6.023×10^{23} (Avogadro's number) molecules of that species. Molecular weight is the sum of the atomic weights of the atoms that constitute a molecule of the compound (same as molar mass); units are of the form kg/kmol, g/mol, or lb/lbmol. Molecular weight is the conversion factor that relates the mass and the number of moles of a quantity of a substance.

5 Compositions of Streams

(1) Mass Fraction and Mole Fraction

Process streams occasionally contain one substance. More often they consist of mixtures of liquids or gases, or solutions of one or more solutes in a liquid solvent. The following terms may be used to define the composition of a mixture of substances, including a species *A*:

$$\text{Mass fraction}: x_{\text{A}} = \frac{\text{mass of } A}{\text{total mass}}$$

$$\text{Mole fraction}: y_{\text{A}} = \frac{\text{mole of } A}{\text{total number of moles}}$$

Mass fractions can be converted to mole fractions or vice versa by assuming a basis of calculation. Remember mass and mole fractions are unitless.

(2) Concentration

Concentrations can be expressed in many ways: weight/weight fraction(w/w), weight/volume fraction(w/v), molar concentration(M), and mole fraction. The weight/weight concentration is the weight of the solute divided by the total weight of the solution, and this is the fractional form of the percentage composition by weight. The weight/volume concentration is the weight of solute divided by the total volume of the solution. The molar concentration is the number of moles of the solute, expressed in moles, divided by the volume of the solution. The mole fraction is the ratio of the number of moles of the solute to the total number of moles of all species present in the solution.

6 Pressure Measurement

Pressure is the ratio of force perpendicular to the area on which the force acts. Pressure units are force units divided by area units (e.g., N/m^2 or Pascal (Pa), dyn/cm^2, and $lbf/in.^2$ or psi). The pressure at the base of a vertical column of fluid (nonmoving) of density ρ and height h, called the hydrostatic pressure, is given by:

$$P=P_o+\rho gh$$

where P_o is the pressure exerted on the top of the column, g is the acceleration of gravity.

The earth atmosphere can be considered as a column of fluid with zero pressure at the top. The fluid pressure at the base of this column (e.g., at sea level) is the atmospheric pressure, P_{atm}.

The **absolute pressure**, P_{abs}, of a fluid is the pressure relative to a perfect vacuum (P=0). Many pressure measurement devices report the gauge pressure of the fluid, which is the pressure of the fluid relative to atmospheric pressure (reference pressure). A gauge pressure of 0 indicates that the absolute pressure of the fluid is equal to the atmospheric pressure. The relationship for converting between absolute and gauge pressure is given by:

$$P_{absolute}=P_{gauge}+P_{atmospheric}$$

Pressure is defined as force per unit area. Pressure can be expressed as a relative pressure or absolute pressure. The units of pressure are:

$$\text{SI unit}: \frac{\text{N}}{\text{m}^2}(\text{or Pa})$$

$$\text{CGS unit}: \frac{\text{dyn}}{\text{cm}^2}$$

$$\text{AS unit}: \frac{\text{lbf}}{\text{in.}^2}(\text{or psi})$$

Chemical engineers are most often interested in pressures that are caused by a fluid. If a fluid is flowing through a horizontal pipe and a leak develops, a force must be applied over the area of the hole that causes the leak to stop the fluid from leaking. This pressure is called the fluid pressure (the force applied must be divided by the area of the hole). This is schematically shown in Figure 1.

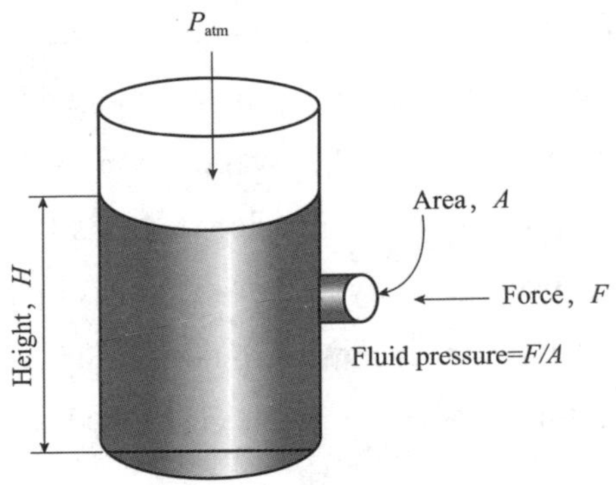

Figure 1 Force on a side hole in an open tank

(a) Atmospheric pressure, P_{atm}, is the pressure caused by the weight of the earth's atmosphere. Often atmospheric pressure is called barometric pressure.

(b) Absolute pressure, P_{abs}, is the total pressure. An absolute pressure of 0.0 is a perfect vacuum. Absolute pressure must be used in all calculations, unless a pressure difference is used.

(c) **Gauge** pressure, P_{gauge}, is pressure relative to atmospheric pressure.

(d) Vacuum pressure, P_{vac}, is a gauge pressure that is below atmospheric pressure. It is used so that a positive number can be reported.

7 Material and Energy Balances

Material and energy balances are very important in the chemical process industries. Material balances can be used to describe material quantities as they pass through a process operation (system). Such balances are statements on the conservation of mass. If no accumulation and generation occur in the system, what goes into a process is equal to what comes out. Material and energy balances can be simple; however, sometimes they can be very complicated. In all cases, the basic approach is the same. A balance on a conserved quantity (i.e., mass or energy) in a system may be written as:

$$\text{Accumulation}=(\text{in} - \text{out}) + (\text{generation} - \text{consumption})$$

Working with simpler systems, such as individual unit operations, is a practical approach to apply the balance to more complicated situations. The increasing availability of computers has meant that very complex mass and energy balances can be set up and manipulated quite readily and, therefore, used in everyday process management to maximize product yields and minimize costs.

Proper understanding and mastering of material and energy balances is critical to the chemical engineering profession. The formulation of the balances on a process system has far-reaching implications. It provides a basis for modeling systems that would be difficult and expensive to study in the lab, assists in the synthesis of chemical processes and evaluation of design processes, guides in the analysis of such systems, and provides a basis for estimating the economic costs and benefits of a process.

Selected from Chapter 1 of "*Principles of Chemical Engineering Processes: Material and Energy Balance* (2nd edition)", Nayef Ghasem, Redhouane Henda. CRC Press, Taylor & Francis Gro up, 2015.

Notes

Large-scale: *adjective.* (1) involving large numbers or a large area; extensive. (2) (of a map or model) made to a scale large enough to show certain features in detail.

Discipline: (1) *noun.* the practice of training people to obey rules or a code of behavior, using punishment to correct disobedience. (2) *noun.* a branch of knowledge, typically one studied in higher education. (3) *verb.* train (someone) to obey rules or a code of behavior, using punishment to correct disobedience.

Scale-up: *noun.* the action of increasing in size or number, expansion.

Optimization: *noun.* the action of making the best or most effective use of a situation or resource.

Transform: *verb.* make a thorough or dramatic change in the form, appearance, or character of. *noun.* the product of a transformation.

Dimension: *noun.* (1) a measurable extent of some kind, such as length, breadth, depth, or height. (2) an aspect or feature of a situation, problem, or thing. *verb.* form or shape (something) to particular measurements.

Significant figures: *noun.* each of the digits of a number that are used to express it to the required degree of accuracy, starting from the first nonzero digit.

Absolute pressure: When pressure is measured relative to a perfect vacuum, it is called absolute pressure.

Questions

(1)What are the definitions of chemical engineering? Which one do you believe is the most appropriate?

(2)What are the duties and goals of chemical engineers?

(3)Please give an example to explain why both values and units are required to measure or count quantities.

(4)How many SI units do you know? Please list the units.

(5)What is the definition of significant figure? How many significant figures do these values have?

(a) 0.2500 (b) 52.15 (c) 0.250×10^5

(6)What is molecular weight? Please check the molecular weight of these chemicals listed below.

(a) H_2SO_4 (b) C_6H_6 (c) $C_6H_{12}O_6$ (d) $C_{16}H_{36}FN$

(7)Why material and energy balances are extremely important in industrial processes?

Part B Reading for Innovation

Introduction to Chemical Engineering Design

1 Nature of Design

This section is a general discussion of the design process. The subject is chemical engineering design, but the methodology described in this section applies equally to other branches of engineering. Chemical engineering has consistently been one of the highest-paid engineering professions. There is a demand for chemical engineers in many sectors of industry, including the traditional process industries: chemicals, polymers, fuels, foods, pharmaceuticals, and paper, as well as other sectors such as electronic materials and devices, consumer products, mining and metals extraction, biomedical implants, and power generation. The reason that companies in such a diverse range of industries value chemical engineers so highly is the following:

Starting from a vaguely defined problem statement such as a customer need or a set of experimental results, chemical engineers can develop an understanding of the important underlying physical science relevant to the problem and use this understanding to create a plan of action and set of detailed specifications, which if implemented, will lead to a predicted financial outcome.

The creation of plans and **specifications** and the prediction of the financial outcome if the plans are implemented is the activity of chemical engineering design. Design is a creative activity, and as such can be one of the most rewarding and satisfying activities undertaken by an engineer. The design does not exist at the start of the project. The designer begins with a specific objective or customer need in mind, and by developing and evaluating possible designs, arrives at the best way of achieving that objective, be it a better chair, a new bridge, or for the chemical engineer, a new chemical product or production process.

When considering possible ways of achieving the objective the designer will be constrained by many factors, which will narrow down the number of possible designs. There will rarely be just one possible solution to the problem, just one design. Several alternative ways of meeting the objective will normally be possible, even several best designs, depending on the nature of the constraints.

These constraints on the possible solutions to a problem in design arise in many ways. Some **constraints** will be fixed and invariable, such as those that arise from physical laws, government regulations, and engineering standards. Others will be less rigid and can be relaxed by the designer as part of the general strategy for seeking the best design. The constraints that are outside the designer's influence can be termed the external constraints. These set the outer boundary of possible designs, as shown in Figure 1. Within this **boundary** there will be a number of plausible designs bounded by the other constraints, the internal constraints, over which the designer has some control; this can include choice of process, choice of process conditions, materials, and equipment.

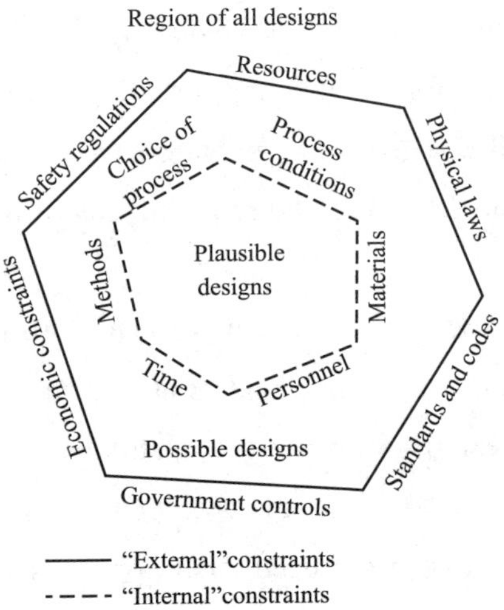

Figure 1 Design constraints

Time will also be a constraint. The time available for completion of a design will usually limit the number of alternative designs that can be considered. The stages in the development of a design, from the initial identification of the objective to the final design, are shown diagrammatically in Figure 2. Each stage is discussed in the following sections. Figure 2 shows design as an iterative procedure. As the design develops, the designer will become aware of more possibilities and more constraints and will be constantly seeking new data and evaluating possible design solutions.

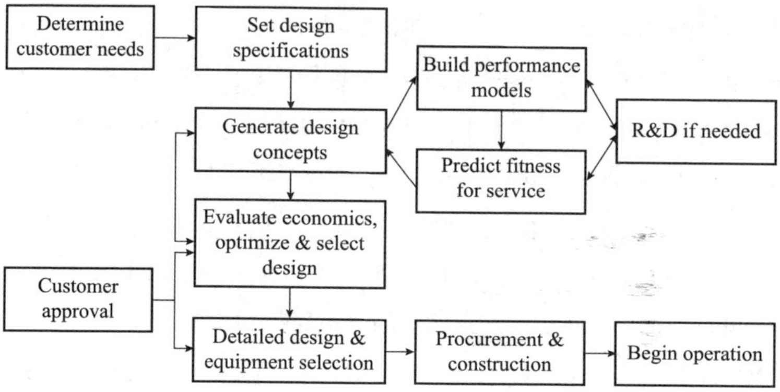

Figure 2 The design process

2 The Design Objective

All design starts with a perceived need. In the design of a chemical product or process, the need is the public need for the product, creating a commercial opportunity, as foreseen by the sales and marketing organization. Within this overall objective the designer will recognize subobjectives: the requirements of the various units that make up the overall process.

Before starting work, the designer should obtain as complete, and as **unambiguous**, a statement of the requirements as possible. If the requirement (need) arises from outside the design group, from a customer or from another department, then the designer will have to elucidate the real requirements through discussion. It is important to distinguish between the needs that are "must-haves" and those that are "should-haves." The "should-haves" are those parts of the initial specification that may be thought desirable but that can be relaxed if necessary as the design develops. For example, a particular product specification may be considered desirable by the sales department but may be difficult and costly to obtain, and some relaxation of the specification may be possible, producing a saleable but cheaper product. Whenever possible, the designer should always question the design requirements (the project and equipment specifications) and keep them under review as the design progresses. It is important for the design engineer to work closely with the sales or marketing department or with the customer directly to have as clear as possible an understanding of the customer's needs.

When writing specifications for others, such as for the mechanical design or purchase of a piece of equipment, the design engineer should be aware of the restrictions (constraints) that are being placed on other designers. A well-thought-out, comprehensive specification of the requirements for a piece of equipment defines the external constraints within which the other designers must work.

3 Setting the design basis

The most important step in starting a process design is translating the customer need into a design basis. The design basis is a more precise statement of the problem that is to be solved. It will normally include the production rate and purity specifications of the main product, together with information on constraints that will influence the design, such as:

(1) The system of units to be used.

(2) The national, local, or company design codes that must be followed.

(3) Details of raw materials that are available.

(4) Information on potential sites where the plant might be located, including climate data, seismic conditions, and infrastructure availability.

(5) Information on the conditions, availability, and price of utility services such as fuel gas, steam, cooling water, process air, process water, and electricity that will be needed to run the process.

The design basis must be clearly defined before design can begin. If the design is carried out for a client, then the design basis should be reviewed with the client at the start of the project. Most companies use standard forms or questionnaires to capture design basis information.

4 The Organization of a Chemical Engineering Project

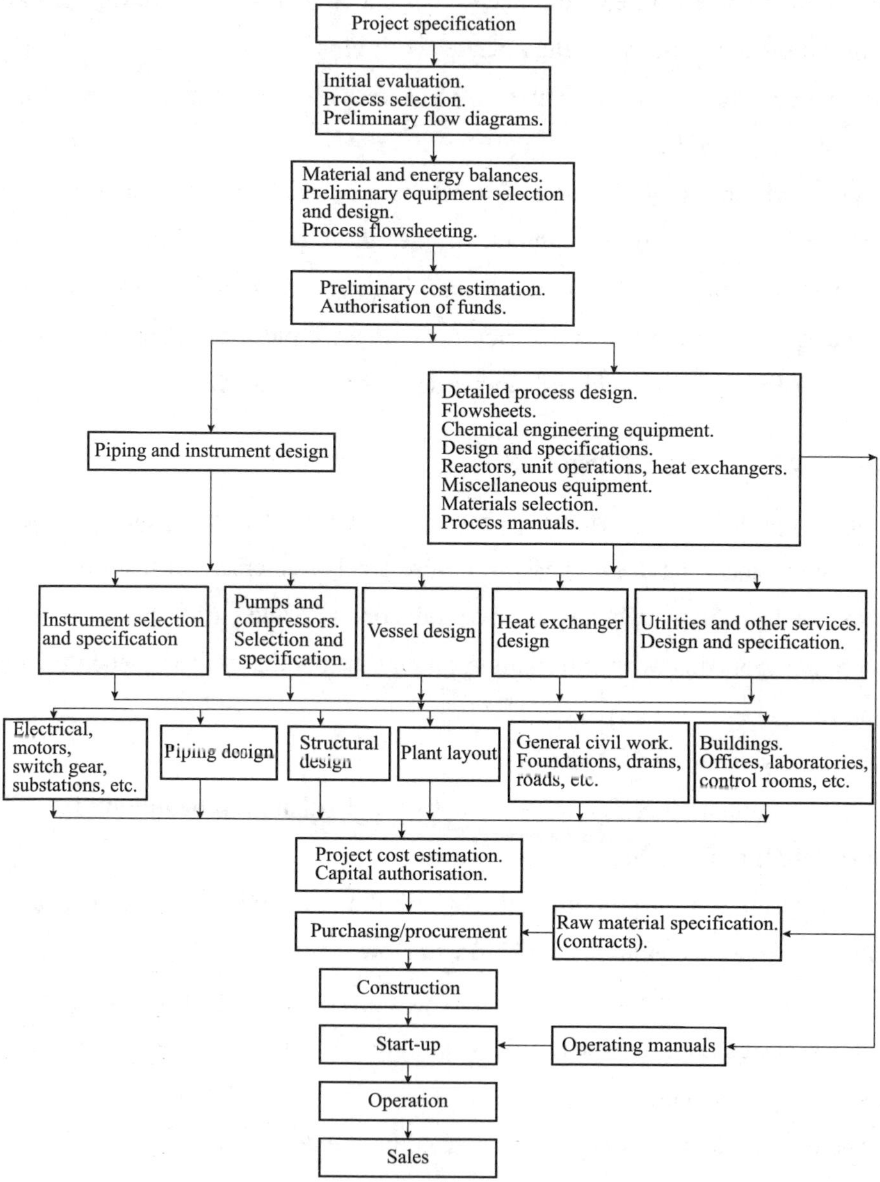

Figure 3 The structure of a chemical engineering project

The design work required in the engineering of a chemical manufacturing process can be divided into two broad phases.

Phase 1: Process design, which covers the steps from the initial selection of the process to be used through to the issuing of the process **flowsheets** and includes the selection, specification, and chemical engineering design of equipment. In a typical organization, this phase is the responsibility of the process design group, and the work is mainly done by chemical engineers. The process design group may also be responsible for the preparation of the piping and instrumentation diagrams.

Phase 2: Plant design, including the detailed mechanical design of equipment; the structural, civil, and electrical design; and the specification and design of the ancillary services. These activities will be the responsibility of specialist design groups, having expertise in the whole range of engineering disciplines. Other specialist groups will be responsible for cost estimation and the purchase and procurement of equipment and materials. The sequence of steps in the design, construction, and **start-up** of a typical chemical process plant is shown diagrammatically in Figure 3, and the organization of a typical project group is shown in Figure 4. Each step in the design process will not be as neatly separated from the others as is indicated in Figure 3, nor will the sequence of events be as clearly defined. There will be a constant interchange of information between the various design sections as the design develops, but it is clear that some steps in a design must be largely completed before others can be started.

A project manager, often a chemical engineer by training, is usually responsible for the coordination of the project, as shown in Figure 4. The project design should start with a clear specification defining the product, capacity, raw materials, process, and site location. If the project is based on an established process and product, a full specification can be drawn up at the start of the project. For a new product, the specification will be developed from an economic evaluation of possible processes, based on laboratory research, pilot plant tests, and product market research. Some of the larger chemical manufacturing companies have their own project design organizations and carry out the whole project design and engineering, and possibly construction, within their own organization. More usually, the design and construction, and possibly assistance with start-up, are subcontracted to one of the international EPC (Engineering, Procurement and Construction) firms.

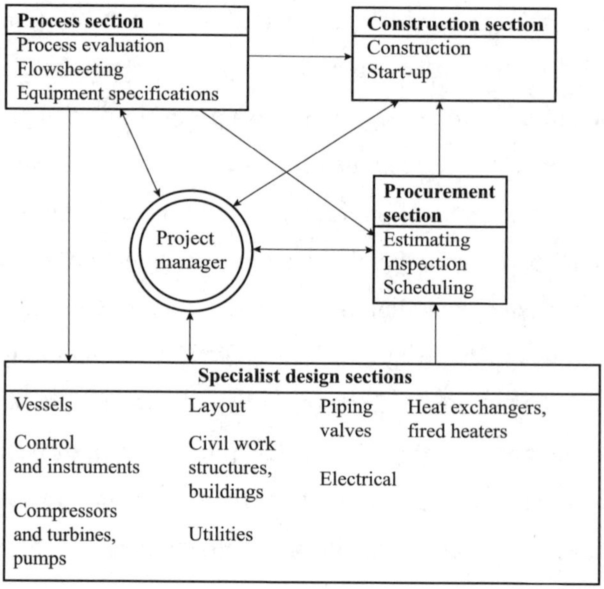

Figure 4 Project organization

The technical “know-how” for the process could come from the operating company or could be licensed from the contractor or a technology vendor. The operating company, technology provider, and contractor will work closely together throughout all stages of the project. On many modern projects, the operating company may well be a joint venture between several companies. The project may be carried out between companies based in different parts of the world. Good teamwork, communications, and project management are therefore critically important in ensuring that the project is executed successfully.

5 Design Factors

Design is an inexact art; errors and uncertainties arise from uncertainties in the design data available and in the approximations necessary in design calculations. Experienced designers include a degree of overdesign known as a design factor, design margin, or safety factor to ensure that the design that is built meets product specifications and operates safely.

In mechanical and structural design, the design factors that are used to allow for uncertainties in material properties, design methods, fabrication, and operating loads are well established. For example, a factor of around 4 on the tensile strength, or about 2.5 on the 0.1% proof stress, is normally used in general structural design. The recommended design factors are set out in the codes and standards.

Design factors are also applied in process design to give some tolerance in the design. For example, the process stream average flows calculated from material balances are usually increased by a factor, typically 10%, to give some flexibility in process operation. This factor will set the maximum flows for equipment, instrumentation, and piping design. Where design factors are introduced to give some contingency in a process design, they should be agreed within the project organization and clearly stated in the project documents (drawings, calculation sheets, and manuals). If this is not done, there is a danger that each of the specialist design groups will add its own "factor of safety", resulting in gross and unnecessary overdesign. Companies often specify design factors in their design manuals.

When selecting the design factor, a balance has to be made between the desire to make sure the design is adequate and the need to design to tight margins to remain competitive. Greater uncertainty in the design methods and data requires the use of bigger design factors.

Selected from Chapter 1 of "*Chemical Engineering Design: Principles, Practice and Economics of Plant and Process Design* (third edition)", Gavin Towler Ray Sinnott, Elsevier, 2022.

Important Words and Phrases

Specification: *noun.* (1) an act of describing or identifying something precisely or of stating a precise requirement. (2) a detailed description of the design and materials used to make something. (3) a standard of workmanship, materials, etc., required to be met in a piece of work. (4) a description of an invention accompanying an application for a patent.

Constraints: *noun.* (1) a limitation or restriction. (2) stiffness of manner and inhibition in relations between people.

Boundary: *noun.* (1) a line that marks the limits of an area; a dividing line. (2) a limit of a subject or sphere of activity.

Unambiguous: *adjective.* not open to more than one interpretation.

Flowsheets: *noun.* a diagram of the sequence of movements or actions of people or things involved in a complex system or activity.

Start-up: *noun.* the action or process of setting something in motion.

Questions

(1) What are the definition and goal of chemical engineering design?

(2) Please name the constraints that might be considered when you design a chemical reactor.

(3) When setting the design objective, what factors should be considered?

(4) What design basis need to be figured out?

(5) There are two broad phases of a chemical manufacturing process design, process design and plant design. What is the main goal of process design and plant design respectively?

(6) What is the definition of design factor?

(7) Design factors are applied in process design to give some tolerance in the design. However, the design factors must be specified in the design manual. Please give explanation to this requirement.

Part C Reading for Fun

General Procedures for Reactor Design

The design of a chemical reactor should not be carried out in isolation from overall process design. It will usually be necessary to determine optimal conditions of reactor conversion, selectivity, and yield of by-products. Optimization of the reactor performance thus requires a design of the process separations and heat recovery, as well as costing and economic analysis. Although the reactor itself is usually a relatively small fraction of the process fixed capital cost, the reactor performance can have a very significant impact on capital and operating costs elsewhere in the process.

The design of a reactor should therefore begin with an overall process design to set rough targets for yield and selectivity. At this stage, there may be very little reaction kinetics information available; indeed, it is very unlikely that there will be sufficient information to develop a model that accurately predicts the effect of feed contaminants and the formation of by-products. All that the designer needs at this point is a rough idea of what yields are feasible in a given **residence time** or **space velocity**; a few good data points from the laboratory are usually sufficient to get started.

The overall procedure for reactor design is given in Figure 1. The steps in this procedure are described next. During several of the steps it may be necessary for the design engineer to interact with researchers to collect additional data, as indicated in the figure.

Step 1. Collect required data

Reactor design always requires more experimental input than the design of other process equipment. Some required parameters, such as heats of reaction and phase equilibrium constants, can be estimated using commercial process simulation models, while some parameters, such as diffusion coefficients and heat and mass transfer coefficients, can be predicted using correlations from the literature; however, almost all reaction rate constants must be measured experimentally.

The collection of reaction engineering data is often an iterative process. As the designer and researchers gain more insight into reactor performance, they may decide to operate the reactor under different conditions or may become aware of different by-products or feed contaminants that affect reaction rates. It is unlikely that the initial experiments or literature search will provide comprehensive information that covers all possible reaction conditions. It will thus be necessary to return to the laboratory and collect more data in the new region of interest at subsequent stages in the reactor design.

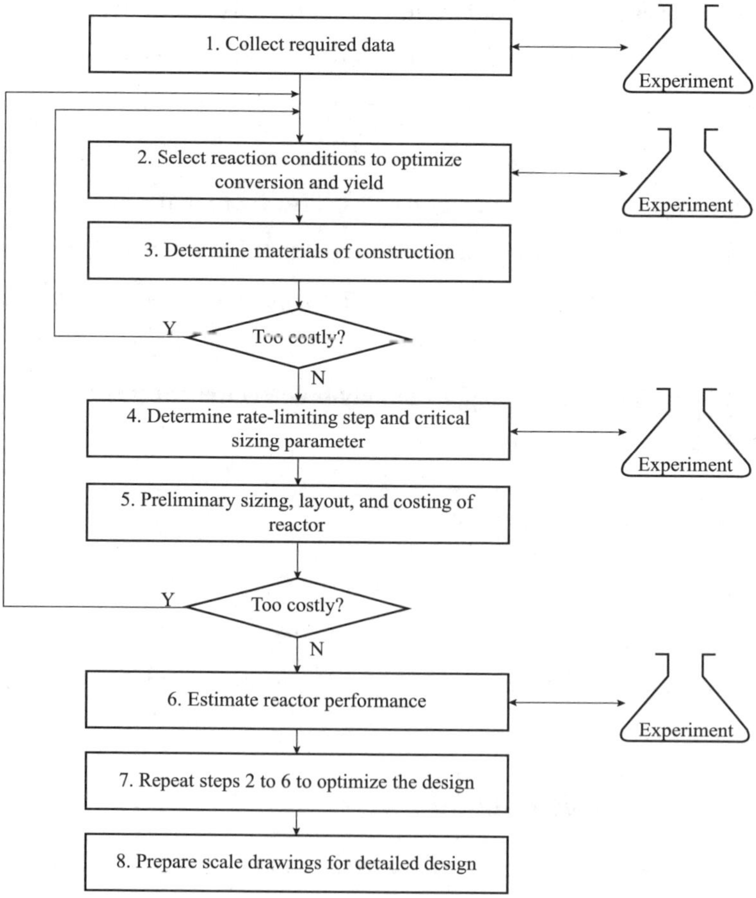

Figure 1 General procedure for reactor design

Step 2. Select reaction conditions

The reaction conditions should be chosen to optimize the reactor conversion, selectivity, and yield, within a reasonable reactor cost and allowing for a safe and controllable reactor design. The choice of reactor conditions is probably the most critical step in reactor design. Some of the conditions, such as the presence of contaminants in recycle streams, may be determined by other process considerations. The selection of reaction conditions will often govern the selection of reactor type. For example, if the reaction conditions cause all the reagents and products to be in the vapor phase, a stirred tank reactor will not be applicable.

Step 3. Determine materials of construction

When the reaction conditions have been specified, the designer can make a preliminary analysis of the materials of construction for the reactor. If the temperature, the pressure, or the presence of a particular component indicates that the reactor will need to be made from a costly alloy, the designer may be tempted to go back to the research team and try to identify a new set of reactor conditions.

Step 4. Determine the rate limiting step and critical sizing parameters of the reactor

In this step the skills and knowledge of reaction engineering are deployed in elucidating the key parameters that determine the extent of reaction under the conditions previously specified. An experimental plan must be carried out under a broad enough range of conditions to determine what physical process limits the extent of reaction. In general, the reaction rate is usually limited by one of the following fundamental processes:

- Intrinsic kinetics: The rate of the reaction itself. In multistep reactions there will usually be one slowest step that governs the overall rate.
- Mass transfer rate: Mass transfer can be particularly important in multiphase reactions, where reagents must transfer between phases, and in reactions that use porous heterogeneous catalysts, where reagents and products must diffuse in and out of catalyst pores. Often, the measured rate of reaction will be confounded with the rate of mass transfer, and careful experimentation will be needed to separate the two effects

and determine which is rate limiting for scale-up.

• Heat transfer rate: If a reaction is **endothermic**, then the temperature of the reaction mixture will decrease unless heat is added. The rate of heat addition can become the governing parameter in reactor sizing, and the reaction may even need to be carried out in a heat transfer device such as a heat exchanger or fired heater.

• Feed addition rate: If the reaction is exothermic, is very rapid, or requires a very low concentration of one of the components for optimal yield, the designer may choose to “starve” the reaction by limiting the feed rate of one of the reagents. Starving a reaction can be a good reactor control strategy for smaller reactors, but can be difficult to accomplish in larger reactors where uniform mixing is more difficult.

• Mixing rate: For very fast reactions, the time taken to mix the reagents can be the limiting step. If the reaction yield is sensitive to feed concentrations, the rate of mixing must be fast enough to ensure that the desired concentration is achieved before reaction occurs (i.e., the reactor must be designed so that mixing is no longer rate limiting).

Step 5. Preliminary sizing, layout, and costing of reactor

Once the sizing parameter of the reactor has been obtained, the designer can estimate the reactor volume, catalyst volume, etc., that will be required at the design rate. The volume estimated is only the active reacting volume, and the reactor layout must also consider the following factors that may add to the volume required for the reactor **vessel**:

• Additional space needed for any internal heat transfer devices such as coils, stab-in bundles, quench zones, etc.

• Additional space needed for **vapor–liquid distribution**, spargers, vapor-liquid segregation or redistribution.

• For liquid-phase stirred tank reactors, some inert vapor space should always be allowed so that the vessel is not filled with liquid. This makes pressure control much easier and reduces the possibility of hydraulic overpressure due to minor changes in flow during operation. A stirred tank reactor should not be designed to operate more than 90% filled, and 65% to 75% is a better design guideline. If foaming can occur, then the level should be at the lower end of the range to allow space for a foam breaker.

• For packed or moving beds, space needed for inert catalyst support materials or support grids.

• For fluidized beds, space needed for fluid distribution grids, fluid-solid segregation, cyclones, transfer lines, allowance for bubble volume, etc.

Step 6. Estimate reactor performance

It is important to confirm that the reactor design will actually achieve the target conversion and selectivities for main products and by-products. Unfortunately, it is often impossible to be fully satisfied without building and testing a full-scale reactor, which is a very expensive proposition.

Historically, chemical companies would go through multiple steps of pilot plant scale-up to validate their reactor designs. Successive pilot plants would be built with the scale of production increasing by one or two orders of magnitude each time. This gradual approach to scale-up was expensive and delayed the market launch of new products and process technologies.

Step 7. Optimize the design

If necessary, Steps 2 to 6 can be repeated to optimize the design. Optimization to reduce the reactor cost is usually a waste of time because the cost of the reactor is typically a relatively small fraction of the total capital cost; however, if the reactor validation experiments showed the presence of unexpected components or showed different selectivities than were found at smaller scale, it will be necessary to reevaluate the overall process optimization and confirm that the target conversion, yields, and selectivities still apply.

Step 8. Prepare scale drawings for detailed design

The final step for the chemical engineer is to make an effective hand-off of the design to the specialized mechanical engineering designer or design team. Because most reactors require substantial internals and ancillary equipment such as agitators, heat transfer coils, etc., the mechanical design of the vessel is rarely trivial. The mechanical engineer needs a complete specification of the vessel and internals, so as to be able to make detailed designs of nozzles, reinforcing rings, internal and external support brackets, and other features. The best means for transmitting all of the required specifications is usually for the reactor designer and the mechanical designer to collaborate on producing scale drawings that show all of the necessary features. Often the mechanical engineer will be involved throughout the design process and will already have a good understanding of the reactor requirements.

Selected from Chapter 10 of "*Chemical Engineering Design: Principles, Practice and Economics of Plant and Process Design* (third edition)", Gavin Towler Ray Sinnott, Elsevier, 2022.

Heat Exchangers

Different heat transfer applications require different types of hardware and different configurations of heat transfer equipment. The attempt to match the heat transfer hardware to the heat transfer requirements within the specified constraints has resulted in numerous types of innovative heat exchanger designs.

The simplest type of heat exchanger consists of two concentric pipes of different diameters, as shown in Figure 1, called the double-pipe heat exchanger. One fluid in a double-pipe heat exchanger flows through the smaller pipe while the other fluid flows through the annular space between the two pipes. Two types of flow arrangement are possible in a double-pipe heat exchanger: in parallel flow, both the hot and cold fluids enter the heat exchanger at the same end and move in the same direction. In counter flow, on the other hand, the hot and cold fluids enter the heat exchanger at opposite ends and flow in opposite directions.

Another type of heat exchanger, which is specifically designed to realize a large heat transfer surface area per unit volume, is the compact heat exchanger. The ratio of the heat transfer surface area of a heat exchanger to its volume is called the area density β. A heat exchanger with β>700m^2/m^3 (or 200 ft^2/ft^3) is classified as being compact. Examples of compact heat exchangers are car radiators (β<1000m^2/m^3),

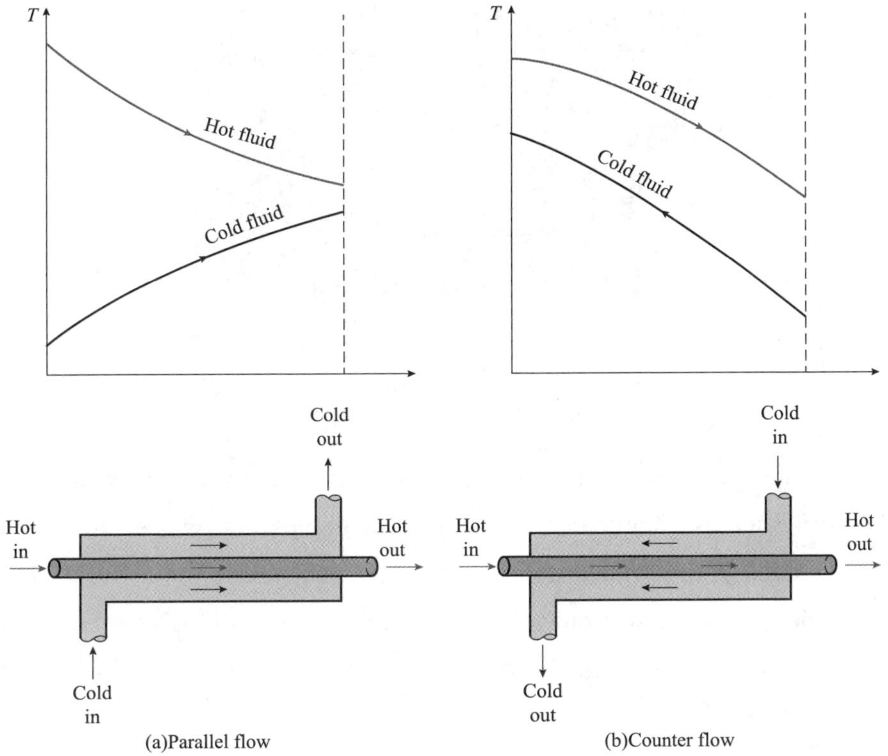

Figure 1 Different flow regimes and associated temperature profiles in a double-pipe heat exchanger

glass-ceramic gas turbine heat exchangers (β<6000m^2/m^3), the regenerator of a Stirling engine (β<15000m^2/m^3), and the human lung (β<20000m^2/m^3). The flow passages in these compact heat exchangers are usually small and the flow can be considered to be laminar. Compact heat exchangers enable us to achieve high heat transfer rates between two fluids in a small volume, and they are commonly used in applications with strict limitations on the weight and volume of heat exchangers (Figure 2).

The large surface area in compact heat exchangers is obtained by attaching closely spaced thin plate or corrugated fins to the walls separating the two fluids. Compact heat exchangers are commonly used in gas-to-gas and gas-to-liquid (or liquid-to-gas) heat exchangers to counteract the low heat transfer coefficient associated with gas flow with increased surface area. In a car radiator, which is a water-to-air compact heat exchanger, for example, it is no surprise that fins are attached to the air side of the tube surface.

Figure 2　A gas-to-liquid compact heat exchanger for a residential air-conditioning system

Another example of compact heat exchangers widely used today in industrial applications such as chemical processing, fuel processing, waste heat recovery, and refrigeration is the printed circuit heat exchanger referred to as PCHE (Figure 3). PCHE derives its name from the manufacturing process in which the flat metal plates that form the core of this heat exchanger are chemically etched to a depth of 1 to 3 mm. The chemically etched plates are then stacked together and joined by diffusion bonding. There may be different etching patterns on the plates to make the heat exchanger as a parallel, counter or cross flow heat exchanger. Furthermore, Researchers reported that unlike conventional heat exchangers with straight flow channels, the flow channels in PCHE can be made in zigzag, S-shape or aero foil shape in order to induce flow turbulence and hence enhance the heat transfer coefficient. PCHEs have high surface density typically greater than $2500m^2/m^3$. The PCHEs are usually made from stainless steel, titanium, copper, nickel and nickel alloys and can withstand operating pressures up to 500bar. The major advantages of PCHE are their wide operating temperature range, which is from −250℃ to 900℃, very high heat transfer coefficients, and their size, which is about four to six times smaller and lighter than conventional heat exchangers. One of the major disadvantages of using PCHEs is very high pressure drop and that they require very clean fluid to be passed through them otherwise blockages can occur easily in the fine channel spacing (0.5 to 2mm).

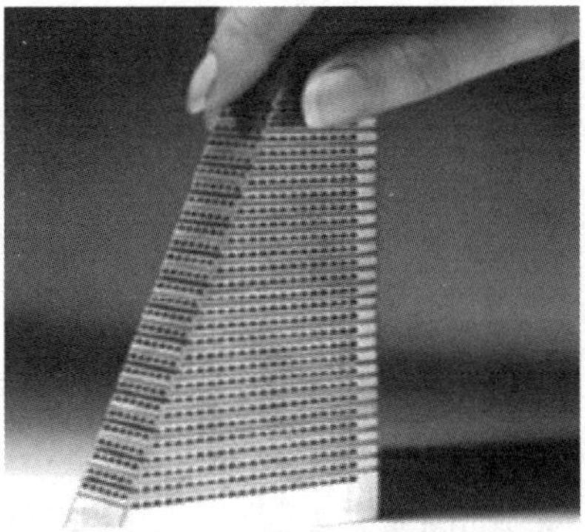

Figure 3 Photograph of a "Core" section from a printed circuit heat exchanger (Heatric, a Meggitt Company, Dorset, UK)

In compact heat exchangers, the two fluids usually move perpendicular to each other, and such flow configuration is called cross-flow. The cross-flow is further classified as unmixed and mixed flow, depending on the flow configuration, as shown in Figure 4. In Figure 4(a) the cross-flow is said to be unmixed since the plate fins force the fluid to flow through a particular interfin spacing and prevent it from moving in the transverse direction (i.e., parallel to the tubes). The cross-flow in Figure 4(b) is said to be mixed since the fluid now is free to move in the transverse direction. Both fluids are unmixed in a car radiator. The presence of mixing in the fluid can have a significant effect on the heat transfer characteristics of the heat exchanger.

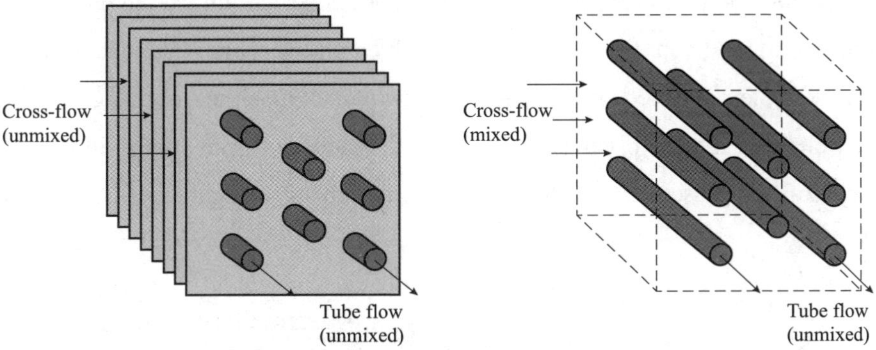

(a) Both fluids unmixed (b) One fluid mixed, one fluid unmixed

Figure 4 Different flow configurations in cross-flow heat exchangers

Perhaps the most common type of heat exchanger in industrial applications is the **shell-and-tube heat exchanger**, shown in Figure 5. Shell-and-tube heat exchangers contain a large number of tubes (sometimes several hundred) packed in a shell with their axes parallel to that of the shell. Heat transfer takes place as one fluid flows inside the tubes while the other fluid flows outside the tubes through the shell. Baffles are commonly placed in the shell to force the shell-side fluid to flow across the shell to enhance heat transfer and to maintain uniform spacing between the tubes. Despite their widespread use, shell-and-tube heat exchangers are not suitable for use in automotive and aircraft applications because of their relatively large size and weight. Note that the tubes in a shell-and-tube heat exchanger open to some large flow areas called headers at both ends of the shell, where the tube-side fluid accumulates before entering the tubes and after leaving them.

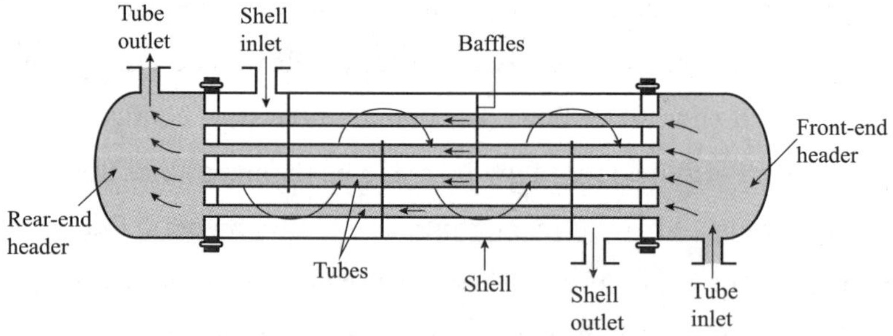

Figure 5 The schematic of a shell-and-tube heat exchanger (one-shell pass and one-tube pass)

Shell-and-tube heat exchangers are further classified according to the number of shell and tube passes involved. Heat exchangers in which all the tubes make one U-turn in the shell, for example, are called one-shell-pass and two-tube-passes heat exchangers. Likewise, a heat exchanger that involves two passes in the shell and four passes in the tubes is called a two-shell-passes and four-tube-passes heat exchanger (Figure 6).

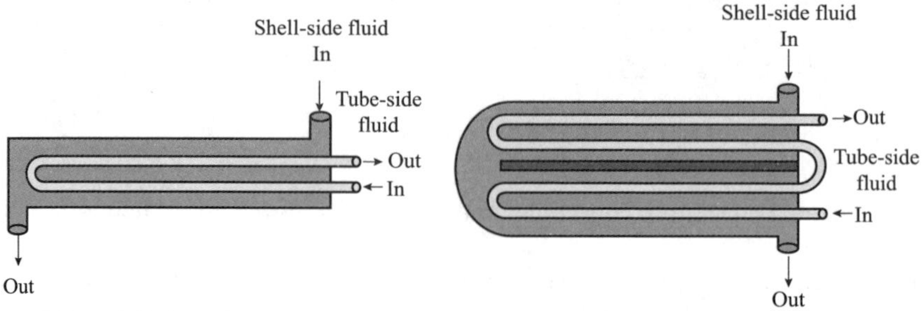

Figure 6 Multipass flow arrangements in shell-and-tube heat exchangers

An innovative type of heat exchanger that has found widespread use is the plate and frame (or just plate) heat exchanger, which consists of a series of plates with corrugated flat flow passages (Figure 7). The hot and cold fluids flow in alternate passages, and thus each cold fluid stream is surrounded by two hot fluid streams, resulting in very effective heat transfer. Also, plate heat exchangers can grow with increasing demand for heat transfer by simply mounting more plates. They are well suited for liquid-to-liquid heat exchange applications, provided that the hot and cold fluid streams are at about the same pressure.

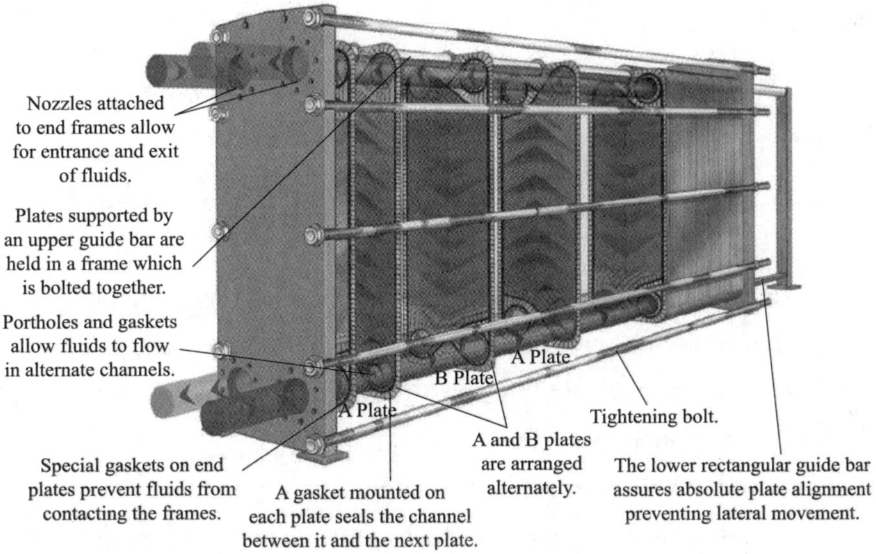

Figure 7 A plate-and-frame liquid-to-liquid heat exchanger

Another type of heat exchanger that involves the alternate passage of the hot and cold fluid streams through the same flow area is the regenerative heat exchanger. The static-type regenerative heat exchanger is basically a porous mass that has a large heat storage capacity, such as a ceramic wire mesh. Hot and cold fluids flow through this porous mass alternatively. Heat is transferred from the hot fluid to the matrix of the regenerator during the flow of the hot fluid, and from the matrix to the cold fluid during the flow of the cold fluid. Thus, the matrix serves as a temporary heat storage medium.

The dynamic-type regenerator involves a rotating drum and continuous flow of the hot and cold fluid through different portions of the drum so that any portion of the drum passes periodically through the hot stream, storing heat, and then through the cold stream, rejecting this stored heat. Again the drum serves as the medium to transport the heat from the hot to the cold fluid stream.

Heat exchangers are often given specific names to reflect the specific application for which they are used. For example, a condenser is a heat exchanger in which one of the fluids is cooled and condenses as it flows through the heat exchanger. A boiler is another heat exchanger in which one of the fluids absorbs heat and vaporizes. A space radiator is a heat exchanger that transfers heat from the hot fluid to the surrounding space by radiation.

Selected from Chapter 11 of "*Heat and Mass Transfer* (5th Edition)", Yunus A. Cengel, Afshin J. Ghajar, McGraw-Hill Education, 2015.

Questions

(1) What is heat exchanger? Why heat exchanger design is required?

(2) Please describe how the heat exchanger as shown in Figure 1 work by your own words.

(3) What is compact heat exchanger? Please give some examples of the application of a compact heat exchanger.

(4) What are the advantages and disadvantages of a compact heat exchanger?

(5) PCHE has been widely used today in industry. What are the advantages of PCHE?

(6) Shell and tube heat exchanger is the most common type of heat exchanger in industrial application. Why do baffles have to be placed in the shell and tube heat exchanger?

(7) How does a plate and frame heat exchanger work? What are the advantages of the plate and frame heat exchanger?

(8) Whether a condenser is one kind of heat exchanger or not? How does a condenser work?

Analogy Between Heat and Mass Transfer

We have spent a considerable amount of time studying heat transfer, and we could spend just as much time studying mass transfer. However, the mechanisms of heat and mass transfer are analogous to each other, and thus we can develop an understanding of mass transfer in a short time with little effort by simply drawing parallels between heat and mass transfer. Establishing those "bridges" between the two seemingly unrelated areas will make it possible to use our heat transfer knowledge to solve mass transfer problems. Alternately, gaining a working knowledge of mass transfer will help us to better understand the heat transfer processes by thinking of heat as a massless substance as they did in the nineteenth century. The short-lived caloric theory of heat is the origin of most heat transfer terminology used today and served its purpose well until it was replaced by the kinetic theory. Mass is, in essence, energy since mass and energy can be converted to each other according to Einstein's formula $E=mc^2$, where c is the speed of light. Therefore, we can look at mass and heat as two different forms of energy and exploit this to advantage without going overboard.

1 Temperature

The driving force for heat transfer is the temperature difference. In contrast, the driving force for mass transfer is the concentration difference. We can view temperature as a measure of "heat concentration" and thus a high temperature region as one that has a high heat concentration (Figure 1). Therefore, both heat and mass are transferred from the more concentrated regions to the less concentrated ones. If there is no temperature difference between two regions, then there is no heat transfer. Likewise, if there is no difference between the concentrations of a species at different parts of a medium, there will be no mass transfer.

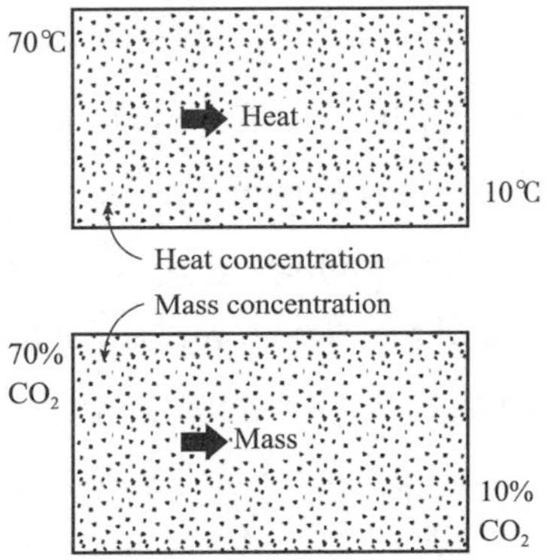

Figure 1 Analogy between heat and mass transfer

2 Conduction

You will recall that heat is transferred by **conduction**, **convection**, and **radiation**. Mass, however, is transferred by conduction (called diffusion) and convection only, and there is no such thing as “mass radiation” (unless there is something Scotty knows that we don’t when he “beams” people to anywhere in space at the speed of light, Figure 2). The rate of heat conduction in a direction x is proportional to the temperature gradient dT/dx in that direction and is expressed by Fourier’s law of heat conduction as:

$$\dot{Q}_{cond} = -kA\frac{dT}{dx}$$

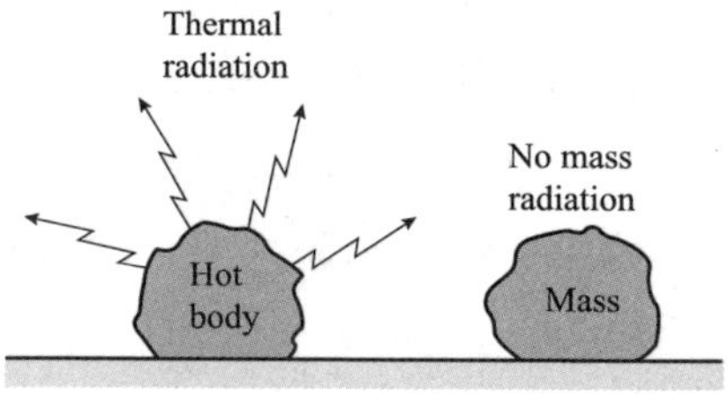

Figure 2 Unlike heat radiation, there is no such thing as mass radiation

where k is the thermal conductivity of the medium and A is the area normal to the direction of heat transfer. Likewise, the rate of mass diffusion $\dot{m}_{diff}$ of a chemical species A in a stationary medium in the direction x is proportional to the concentration gradient dC/dx in that direction and is expressed by Fick’s law of diffusion by (Figure 3):

$$\dot{m}_{diff} = -D_{AB}A\frac{dC_A}{dx}$$

where D_{AB} is the diffusion coefficient (or mass diffusivity) of the species in the mixture and C_A is the concentration of the species in the mixture at that location.

It can be shown that the **differential equations** for both heat conduction and mass diffusion are of the same form. Therefore, the solutions of mass diffusion equations can be obtained from the solutions of corresponding heat conduction equations for the same type of boundary conditions by simply switching the corresponding coefficients and variables.

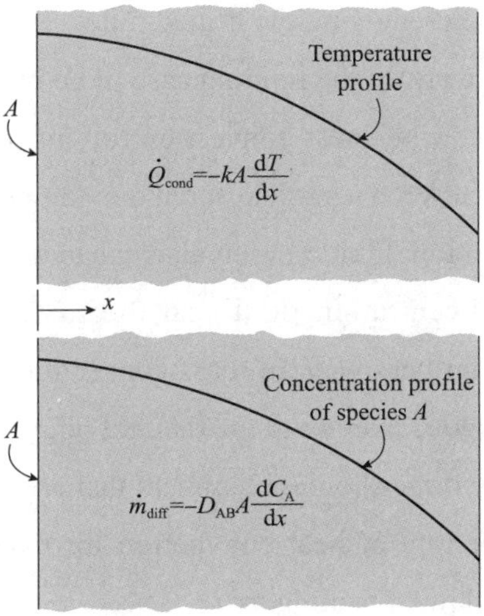

Figure 3 Analogy between heat conduction and mass diffusion

3 Heat Generation

Heat generation refers to the conversion of some form of energy such as electrical, chemical, or nuclear energy into sensible thermal energy in the medium. Heat generation occurs throughout the medium and exhibits itself as a rise in temperature. Similarly, some mass transfer problems involve chemical reactions that occur within the medium and result in the generation of a species throughout. Therefore, species generation is a volumetric phenomenon, and the rate of generation may vary from point to point in the medium. Such reactions that occur within the medium are called homogeneous reactions and are analogous to internal heat generation.

In contrast, some chemical reactions result in the generation of a species at the surface as a result of chemical reactions occurring at the surface due to contact between the medium and the surroundings. This is a surface phenomenon, and as such it needs to be treated as a boundary condition. In mass transfer studies, such reactions are called heterogeneous reactions and are analogous to specified surface heat flux.

4 Convection

You may recall that heat convection is the heat transfer mechanism that involves both heat conduction (molecular diffusion) and bulk fluid motion. Fluid motion enhances heat transfer considerably by removing the heated fluid near the surface and replacing it by the cooler fluid farther away. In the limiting case of no bulk fluid motion, convection reduces to conduction. Likewise, mass convection (or convective mass transfer) is the mass transfer mechanism between a surface and a moving fluid that involves both mass diffusion and bulk fluid motion. Fluid motion also enhances mass transfer considerably by removing the high concentration fluid near the surface and replacing it by the lower concentration fluid farther away. In mass convection, we define a concentration boundary layer in an analogous manner to the thermal boundary.

Layer and define new dimensionless numbers that are counterparts of the Nusselt and Prandtl numbers. The rate of heat convection for external flow was expressed conveniently by Newton's law of cooling as:

$$\dot{Q}_{\mathrm{conv}} = h_{\mathrm{conv}} A_s \left(T_s - T_\infty \right)$$

Where h_{conv} is the heat transfer coefficient, A_s is the surface area, and $T_s–T_\infty$ is the temperature difference across the thermal boundary layer. Likewise, the rate of mass convection can be expressed as (Figure 4):

$$\dot{m}_{\mathrm{mass}} = h_{\mathrm{mass}} A_s \left(C_s - C_\infty \right)$$

Where h_{mass} is the mass transfer coefficient, A_s is the surface area, and $C_s–C_\infty$ is a suitable concentration difference across the concentration boundary layer.

The analogy is valid for low mass transfer rate cases in which the flow rate of species undergoing mass flow is low (under 10 percent) relative to the total flow rate of the liquid or gas mixture.

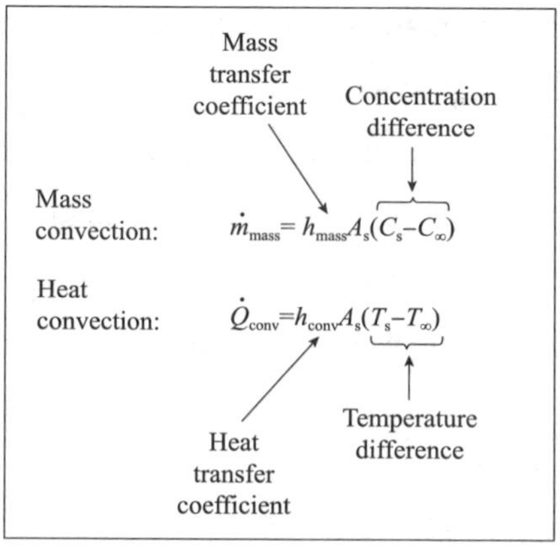

Figure 4　Analogy between convection heat transfer and convection mass transfer

Selected from Chapter 14 of "*Heat and Mass Transfer* (5th Edition)", Yunus A. Cengel, Afshin J. Ghajar, McGraw-Hill Education, 2015.

Questions

(1) What is the driving force for heat transfer? What is the direction of heat transfer?

(2) What is the driving force for mass transfer? What is the direction of mass transfer?

(3) What are the ways of heat transfer?

(4) What are the ways of mass transfer?

(5) Please explain each factor in Fourier's law of heat conduction.

(6) Please explain each factor in Fourier's law of diffusion.

(7) What is heat convection? How can convection be reduced to conduction?

(8) How fluid motion enhances mass transfer?

(9) Please explain each factor in Newton's law of cooling.

Chapter 2 Ethics in Chemical Engineering（化学工程伦理）

© Randy Glasbergen/glasbergen.com

"Have you ever been arrested for shoplifting? We're looking for someone who can help us steal customers from our competition."

Part A Comprehensive Reading

Engineering and Technology

There is an important distinction between engineering and technology, and hence the work of engineers and technologists, that is often overlooked and which requires attention here. Engineering is primarily a **profession** in which activities are carried out by persons for the benefit of other persons. Such activities involve the development or use of technology, but effective engineering does not necessarily involve the use of advanced technology. For example, effective engineering may involve an ingenious but technically simple means of meeting a genuine human need. Indeed, there is always a danger that engineers may become so involved in technical wizardry that the **ethical** dimension, the effect of the technology on others, may be neglected or lost. In contrast, the primary goal of technology and technologists tends to lie more in the direction of the development of artefacts. Furthermore, only a very sophisticated device is likely to win a technology prize, at least in developed countries. This contrast between engineering and technology is not sharp, for technologists may also have a concern for human benefit, but it has significant ethical implications.

Some aspects of this distinction between engineering and technology may be illustrated by considering the example of the bicycle. In the prestigious Reith lectures of 2005, Alec Broers, at that time the President of the Royal Academy of Engineering, provided the evaluation, "I would argue…that most technologies, with the exception of those associated with weaponry, have had hugely beneficial effects for most people". However, Broers was very surprised by a poll that asked the public how they would rank Britain's greatest inventions. Over 50% of 5000 respondents had ranked the bicycle highest. He wrote, "The bicycle is of course an ingenious, practical and sustainable invention, which has brought new opportunities to every stratum of society, and which continues to offer benefits today", but he considered that to put it ahead of inventions such as electricity, jet engines and vaccination was "a profound misunderstanding of the contribution of advanced technologies in our lives".

This assessment is valid, but it should also be noted that although a bicycle is a simple artefact it is an excellent example of effective engineering. Part of its ingenuity lies in its counter-**intuitive** nature: it is surprising that a person can so readily learn to balance and control a single-track vehicle, and mathematical simulation of the skills of bicycle riding is mathematically formidable, if not intractable. A bicycle has a modest cost, is non-polluting, considerably increases the distance that a person can travel in 1 day, and can carry loads of 10 times its weight. Furthermore, the engineering development of bicycles continues. In the developed world this includes the engineering of ingenious designs of lightweight folding bicycles for commuting. In the developing world bicycles are an important form of transport and may also be used to meet local needs in many ingenious ways: to husk corn, to sharpen knives, to charge batteries, to filter water, as an ambulance, to cross rivers, to pump water, to pump out a pit latrine, to wash clothes, and as parts of a windmill.

Misconceptions about engineering and misuse of technology also have influence in areas of human endeavor not usually associated with those disciplines, such as economics and banking. Amartya Sen, the Nobel Prize winning economist and philosopher, has argued that economics has two different origins. The first he identifies as ethics, the good of man in a social context. The second he disparagingly refers to as an "engineering" approach that is concerned primarily with logistical issues, with means rather than ends, seeing human beings in very narrow terms and characterizing social institutions in rather simple form. Sen favors a synthesis of economics and ethics, and argues that such closer contact can be beneficial not only to economics but also to ethics. Though consideration of means is also important, Sen's terminology clearly involves a misunderstanding of engineering. Nevertheless, his work to create a synthesis of economics and ethics should find a parallel in the creation of a synthesis of the technical aspects of engineering and ethics.

A useful guide to an ethical approach to life is the sequence of **awareness**, compassion and action: awareness to recognize those whose well-being we can influence, compassion for those in need, and effective action to remedy those needs. For professional engineers such awareness needs to extend well beyond the purely technical aspects of their work.

For example, consider the link between smoking and lung cancer, which might at first appear to be a medical and social issue with no engineering component. At the start of the twentieth century, lung cancer was such a rare disease that medical professors when confronted with a case sometimes told their students that they might never see another. Cigarettes now cause 1.5 million deaths from cancer each year, a number that is expected to rise to nearly 2 million each year by the 2020s or 2030s, even if consumption rates decline in the interim. The cigarette has been described as "the deadliest artefact in the history of human civilization". Technological advances played a very important part in this change.

Up to the end of the nineteenth century, cigarettes were produced by hand with a skilled worker able to produce about 200 in a shift. Such procedures were enough to satisfy a modest demand. This slow production rate was revolutionized by the invention of a mechanically complex but very efficient cigarette-manufacturing machine, as described in two patents. Each new machine could produce 120000 cigarettes in a day, about a fifth of US consumption at that time. Such technical innovation was followed by huge advertising and marketing campaigns, leading eventually to the present worldwide rate of consumption of about 6 trillion cigarettes each year. Cigarettes were initially promoted as beneficial for health. However, the link to lung cancer was becoming apparent by the 1930s and was clearly established by the 1950s. Cigarette manufacture puts a very low value on human life: "Cigarette companies make about… US$10000 for every million cigarettes purchased. Since there is one death for every million cigarettes sold (or smoked), a tobacco manufacturer will make about US$10000 for every death caused by their products…The value of a human life to a cigarette manufacturer is about US$10000". Many causal factors are involved in continuing cigarette use: economic, political, agricultural and so forth. Many different agents are also involved, including farmers, investors, executives, advertisers and retailers.

Engineers now play a key role in the manufacturing process, including the design, manufacture and operation of modern equipment. The largest factory in operation produces 146 billion cigarettes a year (in 2010) and hence probably causes about 146000 deaths annually. When the technology for rapid cigarette manufacture was invented, the health effects of smoking were not known. However, the engineers now working in the cigarette industry clearly have major ethical issues to address, for they are engaged in what is known to be one of the deadliest of all engineering activities. They may represent only one professional group in the industry, but without their skills the industry could not operate. Those engineers working in the industry may wish to seek alternative employment, for engineering knowledge and skills may be applied for many purposes. Likewise, other engineers seeking employment should ask themselves whether they wish to work in an industry the products of which lead to such dire consequences for so many. Such ethical concerns are especially significant for young engineers, for early employment choices can strongly influence the engineering, and hence ethical, direction of an entire career.

Awareness, compassion and action can be subject to constraints in professional working environments. For example, an official report into the nuclear accident at Fukushima in 2011 found a multitude of errors and instances of willful negligence, but concluded that the fundamental faults lay in the working culture: "our reflexive obedience; our reluctance to question authority; our devotion to 'sticking with the program'; our groupism; our insularity". However, it will also be found that engineers have great opportunities to accept and express ethical responsibility in imaginative ways.

Selected from Chapter 1 of "*Engineering Ethics: Challenges and Opportunities*", W. Richard Bowen, Springer (2014).

Notes

Profession: *noun.* (1) the act of taking the vows of a religious community. (2) an act of openly declaring or publicly claiming a belief, faith, or opinion. (3) an avowed religious faith.

Ethical: *adjective.* (1) relating to moral principles. (2) (of a medicine) legally available only on a doctor's prescription and usually not advertised to the general public.
Intuitive: *adjective.* using or based on what one feels to be true even without conscious reasoning; instinctive.
Awareness: *noun.* knowledge or perception of a situation or fact.

Questions

(1) Are you able to share your personal understanding on "profession"?
(2) How do you think about the emergence of electronic vehicles (EV)?
(3) How to manage the activity with nuclear energy, on both domestic and international levels?

Part B Reading for Innovation

Adapting Educational Experiences for the Chemists of Tomorrow

Technical universities are constantly experimenting with innovative student-centered classroom approaches. We describe **flipped** and linked classroom approaches in the context of our ongoing chemical education theme of catalysis.

The calling to teach the next generation of chemists in East Asia has led us and others to search for new modes of learning that can inspire creativity and self-reliance. Fresh experiences in learning arise from experiments in teaching. If we are not careful, from a student's perspective, the result might be trials and **tribulations**. Creativity in (chemistry) education comes down to different approaches. These forays, some linguistic in nature, could be anything from broad campus-wide initiatives to grassroots efforts. Grand initiatives require heavy cooperation, "branding" and looking to the future. In brief, they require campus management. At technical universities, where developments in science and technology are often prioritized over new educational paradigms, individual professors and educational professionals alike seek to test their ability to create the classroom of the future.

1 A Case Study in Flipped Classroom Learning

Regarding campus-wide efforts, about 10 years ago, Korea Advanced Institute of Science and Technology (KAIST) launched an educational platform called Education 3.0. In this flipped format of learning, students watch pre-recorded online videos at home and then attend weekly face-to-face meetings. Here, they convene at round tables, prepare and deliver presentations, and participate in team-based discussion and learning on themes prescribed by a professor as per the syllabus. Under "3.0", learners performed projects, tasks and experiments together in a more embracing and **interactive** mode than was typically possible in traditional large lecture halls, exemplifying a marked difference from the then status quo. Unsurprisingly, because of this substantial change in classroom style, both professors and students at first found frustration. Amidst the faculty lunch dialogue, a pervasive joke emerged: "you can lecture any way you want, except by lecturing". In response to these frustrations, a new "one-stop shopping" approach was implemented across the campus, offering a tremendous array of alternatives deviating from the heavy reliance on video instruction of a flipped format course. Instead, fresh course plans were devoid of lectures and full of student interaction and collaboration through semester-long project-based and/ or problem-based learning (Education 4.0). This widespread experience of online and blended instruction across the curriculum also facilitated remote learning, allowing the quality of KAIST instruction to be largely unaffected by the COVID-19 pandemic.

2 Chemistry Education

Amidst campus-wide initiatives, relying on great numbers of faculty marching to a proposed administrative drumbeat, there are subtler grassroots efforts to develop education methods that are sometimes **esoteric** but often more specific to a certain field of study.

A substantial amount of the coursework at any technical university involves chemistry in one form or another, such that it lives up to its place as the "central science". In addition, because professors have to get the point across, every such instructor, student and teaching assistant alike is faced with issues relevant to chemistry education whether they know it or not. Faculty members have been active in publishing articles covering topics from aspects of laboratory environment, chemical safety and general chemistry, as well as inter-language resources to help bring about new modes of learning.

New dimensions of chemistry education and related training are being tested all the time. For example, the need to produce clear and compelling **prose** is important in securing research funding and getting papers accepted. To this end, we have been running a publishing club through which professors can share insights and tricks of the trade with students when submitting and responding to reviewer feedback. In addition, manuscript writing and presentation courses were invigorated by taking into account that linguistic challenges are coupled with scientific ones. The English as a Foreign Language Department at KAIST now provides discipline-specific Scientific Writing and Scientific Presentation courses for graduate students to afford them deep insights into the features that distinguish the dissemination of chemistry from that of other subjects.

3 Catalysis as an Example

Research in scientific education and pedagogy allows us to explore how technology can enhance the classroom and how we abstract and visualize chemical concepts. As an example, an international project aims to put a spin on how we teach catalysis and create iconography for catalysts and the field of catalysis. One way to understand catalysis is to understand what ruins an otherwise happily working catalyst. Poisons and inhibitors can begin to shut down catalytic cycles that normally occur smoothly, whether it be in a catalytic converter, an enzyme in our body or an *n*th generation catalyst for industry. New ways to represent catalysts can be deeply explored from a hybrid education-science perspective.

Faculty members who teach science and engineering courses and those who teach communications courses can collaborate, often in globalized ways, to leverage technology enhanced learning to increase instructional depth and discipline specificity in the instruction of scientific communication. Catalysis coverage can be deep yet tailored. Information and communication technologies can afford students access to a greater array of learning content to suit their individual needs, particularly with regard to the instruction of scientific communication. At KAIST, as in many global universities, students have varying levels of experience and skill with their research, with visual and verbal communication in their field, and with the English language in general. The ongoing research described here aims to teach the concept of inhibitors in an overlapping manner between content classes and communications classes; these can be linked by prior cooperation of instructors working in different departments. We hope that our approach will enable students to achieve an enriched understanding of inhibitors and their effects on chemical reactions, as students aim to express, both textually and visually in communication classes, these concepts to be understood by their peers first and a wider audience second.

Developing a new set of icons for catalysis is a long-term project. With computer-generated graphical chemistry representations and creative journal cover artwork appearing daily, as well as new representations of ligands, students can start by first identifying which previous artwork and pictorial renditions they find clearest and most helpful in conveying the necessary message.

4 2021 and Beyond

Ultimately, we are trying to optimize the educational experience in the era following

COVID-19. More than ever, students are involved in student-centered hybrid learning. To alleviate students going vacuously from classroom to classroom, brilliant educational ideas coupled with course content can help retain long-term knowledge for one's life and scientific or business career. Beauty is often in the eye of the beholder in empowering student and professor initiatives alike.

Selected from Nature Reviews Chemistry, 2021, 5, 141. M Fanguy, S Y Lee, D G Churchill.

Important Words and Phrases

Flip: *verb.* (1) turn over with a sudden quick movement. (2) move, push, or throw (something) with a sudden sharp movement. (3) suddenly lose control or become very angry.

Tribulation: *noun.* (1) a cause of great trouble or suffering. (2) a state of great trouble or suffering.

Interactive: *adjective.* (1) (of two people or things) influencing or having an effect on each other. (2) allowing a two-way flow of information between a computer and a computer-user; responding to a user's input.

Esoteric: *adjective.* intended for or likely to be understood by only a small number of people with a specialized knowledge or interest.

Prose: *noun.* the ordinary language people use in speaking or writing.

Questions

(1) Is flipped and linked classroom approach enhancing your learning so far? Any suggestions?

(2) What is the difference between project-based or problem-based learning with lecture hall style study?

(3) This text employed "catalysis" as an example for new ways of teaching. Are you experiencing other unique and interesting learning course in your college career?

(4) In your institute, do faculty members offer such interactive teaching/learning mode in your subject? How?

Part C Reading for Fun

Energy, Poverty and Development

Modern energy sources provide a crucial underlying basis for the promotion of human flourishing. Nevertheless, the link between poverty and energy provision has received much less attention than the links between poverty and the provision of food, clean water, sanitation and shelter. The effects of poor availability of energy are less immediately apparent than the effects of these other basic goods. However, on a global basis, 1.3 billion people are without access to electricity and 2.7 billion are without clean, safe cooking facilities. More than 95% of these are in sub-Saharan Africa or developing Asia and 84% are in rural areas. Sub-Saharan Africa is home to 12% of the global population of which 45% are without access to electricity. Over 1.9 billion people in developing Asia rely on traditional biomass for cooking, including around 840 million in India and 100 million in each of Bangladesh, Indonesia and Pakistan. A further 400 million people, mostly in China, rely on coal. Even in oil-rich Nigeria there are over 90 million people without access to clean, safe cooking facilities. This lack of modern, safe energy sources has many dire consequences. Just to give one example, it is estimated that cooking on open fires in poorly ventilated or inefficient stoves fueled with traditional biomass results every year in sub-Saharan Africa in about 360000 premature child deaths (under 5 years of age) and about 23000 female deaths (over 30 years of age), from pneumonia and chronic obstructive pulmonary diseases respectively. Global premature deaths due to smoke from such biomass currently exceed those from either malaria or tuberculosis.

Hence, it is particularly important to consider the benefits of appropriate access to energy. Provision of energy is an excellent example of how engineering enables human wellbeing and agency. In the context of extreme poverty, such enablement can be considered in terms of the United Nations Millennium Development Goals. These were established at the UN General Assembly Millennium Meeting in 2000 in recognition of the need to speed up poverty alleviation and socio-economic development. Here are some of the contributions that better energy availability can make to achieving these goals:

(1) Eradicate extreme poverty and hunger. Access to reliable energy enables enterprise development; lighting permits income generation beyond daylight hours; machinery increases productivity; energy can often be provided by small-scale locally owned businesses, creating employment; 95% of staple foods (such as rice, grain and green bananas) need cooking before they are eaten and need water for cooking; improved productivity throughout the food chain, including transportation of food to areas of need; reduced post-harvest losses through better preservation; better irrigation and hence improved food yields; better access to clean water; reduction in proportion of income spent on energy.

(2) Achieve universal primary education. Creation of a more child-friendly environment (clean water, sanitation, lighting, space heating/cooling); freeing of children's, and especially girls' time from helping with survival activities (gathering firewood, fetching water); lighting permits home study; lighting in schools allows evening classes and helps retain teachers, especially if their accommodation has electricity; electricity enables access to educational media that allow distance learning; electricity allows use of teaching aids such as projectors and computers; modern energy systems and efficient building design reduce heating/cooling costs.

(3) Promote gender equality and empower women. Freeing of girls' and young women's time from survival activities (gathering firewood, fetching water, cooking inefficiently, crop processing by hand, manual farm work); lighting permits home study and allows evening classes, giving better access to education; street lighting improves women's safety; reliable energy services increase scope for women's enterprises.

(4) Reduce child mortality. Reduction of the indoor air pollution that contributes to respiratory infections that account for up to 20% of child deaths; reduction in the need for gathering and preparing traditional fuels that exposes children to health risks and reduces time spent in child care; modern energy leads to fewer burns, accidents and house fires; provision of nutritious cooked food, space heating and boiled water contribute to better health; cold chain provision allows access to vaccines.

(5) Improve maternal health. Access to better medical facilities for maternal care, including medicine refrigeration, equipment sterilization and operating theatres; reduction in excessive workload and heavy manual labor improves health of pregnant women; improved production and distribution of sex education literature and contraceptives; access to distance medicine; improved provision of cooked food, space heating and boiled water.

(6) Combat HIV/AIDS, malaria and other major diseases. Electricity in health centers enables night availability, helps retain qualified staff and allows equipment use; refrigeration allows vaccine and medicine storage; safe disposal of used hypodermic syringes by incineration prevents re-use and potential further spread of HIV/AIDS; enables development, manufacture and distribution of medicines and vaccines; allows access to health education media.

(7) Ensure environmental sustainability. Increased agricultural productivity is enabled through the use of machinery and irrigation, which reduces the need to expand the quantity of land under cultivation, reducing pressure on the ecosystem; substitution and improved efficiency reduces erosion, reduced soil fertility and desertification due to traditional fuel use; encouragement of better natural resource management and reduced local pollution; facilitates rural non-farm-based enterprise and processing of non-timber forest products.

(8) Develop a global partnership for development. Energy is essential for most of the cooperative international, governmental, non-governmental and commercial activities that can promote development.

Overall, provision of appropriate access to energy can be a very effective means of contributing to human flourishing in developing countries. However, a number of mistakes have been made in past energy projects and policies, including : prioritizing technology development over delivering energy as a service to the poor; failing to understand the links between energy and livelihoods, including in relation to development and poverty reduction strategies; lack of awareness on the part of non-energy specialists of the role energy plays in meeting objectives in sectors such as agriculture, health, education and manufacturing; lack of attention to the energy requirements of community facilities, schools, medical centers and micro-enterprises; overemphasis on energy supply issues in relation to end-use benefits, represented in the funding bias towards large power projects over capacity building at local level to install, operate and maintain small-scale systems; one-directional planning of energy projects, often ignoring the need to involve communities in the planning process, resulting in projects that are not socially acceptable, sustainable or appropriate to the needs of a given location. That is, many past projects have prioritized technology rather than people.

However, approaches are now being proposed that focus first on understanding the effect that availability of energy has on peoples' wellbeing and agency. The provision of energy can then be designed to ameliorate the constraints on such human flourishing. Key features of such a people-centered approach include：

- Reaching beyond technical issues, to deliver energy provision that meets people's needs and priorities. That is, provision of supply should be designed around the benefits that energy can provide.
- Being aware of subjective and social benefits. Access to electricity promotes a sense of inclusion in modernizing processes. Energy for transport and communication allows poor people to maintain contact with their extended family and friends.
- Ensuring that communities have a voice in the decision making process on how to meet their energy needs. That is, planning should include substantial bottom-up participation.
- Working across all sectors (including agriculture, health, education and manufacturing) to integrate energy more fully into development processes from early on. Energy is a vital enabler in meeting the goals of these sectors.
- Working at local, national and international levels, in order to develop pro-poor policies based on real evidence of the impact of energy availability on poor people. Provision of energy needs to be given a priority comparable to those given to provision of food, clean water, sanitation, shelter and medical services.
- Taking a holistic approach to energy across sectors rather than a project-based

approach. Diverse sources of energy are needed, including appropriate use of traditional sources.

- Building a deeper understanding of the links between energy availability and poverty reduction. This should include assessment of the effects of energy availability on rural to urban migration, the effects on empowerment and inclusion (especially for women), and the effects on the reduction of drudgery.

A great variety of technologies are in principle available, but these are only practically useful if they are available locally at an appropriate price. Some are simple, cheap and effective, such as improved stoves and solar water heaters that can be manufactured locally. Local development and manufacture of small capacity off-grid equipment for the provision of electricity can also be effective, such as micro-hydro turbines, small wind turbines and biogas digester systems. These can be very useful in remote locations. At a village or district level, mini grids can provide electricity with loads up to 500kW. For urban areas, large-scale thermal or hydro electricity generation is necessary. Such sources are important for businesses as well as for individuals. For example, many major companies in Nigeria provide their own electricity through diesel generators. The high cost of such energy supply has been a key factor in limiting the competitiveness of Nigerian industrial production and hence of employment.

An assessment of the balance of differing scales for supply of electricity is important. An International Energy Agency (IEA) projection regards grid extension as being the most suitable for urban areas and for about 30% of rural areas, but not cost-effective for more remote rural areas. Therefore, it is projected that 70% of rural areas should be connected either with mini-grids (65% of this share) or with small, stand-alone off-grid supplies (the remaining 35%). Most of the mini-grid and off-grid provision is projected to come from renewables. Such smaller scale provision can ease the difficulty of finding the funds needed for capital investment, though modern renewable technologies tend to have relatively high initial capital costs and relatively low recurrent costs.

The IEA has carried out detailed studies of the costs of providing modern energy access, defined as “a household having reliable and affordable access to clean cooking facilities, a first connection to electricity and then an increasing level of electricity consumption over time". Here clean cooking facilities refers to biogas systems, liquefied petroleum gas (LPG) stoves and advanced biomass stoves that have much lower emissions and higher efficiencies than traditional fires. The initial electricity consumption for rural households was assumed to be 250kW·h per year, and 500kW·h per year for urban households. The IEA considered the period 2010~2030, with electricity consumption of newly connected households rising to 800kW·h per year by 2030, a level seen in much of developing Asia. These are relatively modest amounts compared to present Western levels, though the figures do not include provision to businesses and public buildings such as schools and hospitals. It was estimated that to provide universal modern access at these levels would require investment of US$48 billion each year, on average. This greatly exceeds the estimated US$9.1 billion that was invested globally in extending access to modern energy provision in 2009. However, these are small amounts compared to global weapons sales of US$411 billion and world military expenditure of US$1738 billion in 2011. The IEA estimates that achieving universal access by 2030 would increase global electricity generation by 2.5%. The demand for fossil fuels would be expected to rise by 0.8% and CO_2 emissions would increase by 0.7%, figures it considers trivial in relation to energy security or climate change. The benefits would be a major contribution to social and economic development and a contribution to avoiding 1.5 million premature deaths per year. The provision of such energy resources can be seen as a clear opportunity of professional capabilities for engineers.

Selected from Chapter 5 of "*Engineering Ethics: Challenges and Opportunities*", W. Richard Bowen, Springer (2014).

Dishonesty in Scientific Communities

1 Dishonesty on Campus

Problems for engineering students regarding honesty in engineering can arise even before completing their undergraduate work. Here is a real-life example. John is a co-op student with a summer job with Oil Exploration, Inc., a company that does exploratory contract work for large oil firms. The company drills, tests, and writes advisory reports to clients based on the test results. As an upper-level undergraduate student in petroleum engineering, John is placed in charge of a field team of roustabouts and technicians who test drill at various sites specified by the customer. John has the responsibility of transforming rough field data into succinct reports for the customer. Paul, an old high school friend of John's, is the foreperson of John's team. In fact, Paul was instrumental in getting this high-paying summer job for John.

While reviewing the field data for the last drilling report, John notices that a crucial step was omitted, one that would be impossible to correct without returning to the site and repeating the entire test at great expense to the company. The omitted pumped into the test drill site. The test is important because it provides the data for deciding whether the drill site is worth developing for natural gas protection. Unfortunately, Paul forgot to add the test chemical at the last drill site.

John worries that Paul is likely to lose his job if his mistake comes to light. Paul cannot afford to lose his job at a time when the oil business is slow and his wife is expecting a child. John learns from past company data files that the chemical additive indicates the presence of natural gas in approximately 1 percent of the tests.

Should John withhold from his superiors the information that the test for natural gas was not performed? Given his friendship with Paul and what he knows about Paul's family life, no doubt John feels strong pressure to say nothing. Still, viewed as an ethical problem, there is much more that John needs to take into account. First and foremost, he must realize that, as an employee of Oil Exploration, Inc., he has special responsibilities to his employer that are not grounded in his friendship with

Paul or anyone else. Included in these responsibilities is helping Oil Exploration meet its obligations to its clients. Oil Exploration is expected to present its clients with reliable information. In turn, its clients need to make important decisions on the basis of the information they receive. Both Paul and John have important roles to play in enabling others to succeed in meeting their responsibilities. John is worried that Paul is likely to lose his job if his company finds out about the oversight. Is this worry well-grounded? If John would not fire Paul for his oversight, why does he assume that Oil Expectations management would? Is it because John himself wonders if Paul's oversight, if discovered by management, really would warrant dismissal? In any case, whose call should this be? Would John support the firing of Paul were it not for their friendship? Should John allow friendship to influence his decision? Although attempting to answer these questions may not make it easier for John to decide what to do (quite the contrary), asking them is morally important. What John needs to bear in mind is that he and Paul are working in a professional setting and this means that each has special responsibilities that are wedded to that setting and that are not grounded in their relationship as friends.

A student's experience in engineering school is a training period for his or her professional career. If dishonesty is as detrimental to engineering professionalism as we have suggested, then part of this training should focus on the importance of professional honesty. However, it might be thought that, in general, cheating as a student, especially if one succeeds in getting away with it, is a relatively minor matter. The three students above were unlucky. Their intention, they might plead, was simply to get credit for work that they were confident would turn out well. But they felt they could not wait for the actual results before submitting them for publication. At least, they might have thought, they credited the professor by listing him as lead author. Of course, they probably realized that listing an established researcher as lead author might enhance the chances of the paper being accepted not to mention their chances of getting their nonacademic careers off to a good start.

So, was the students mistake simply one of miscalculation of the odds of ultimate failure in the research itself, of the professor not finding out about their scheme, or of the professor not minding that he was listed as lead author? What if they thought that the academic world is simply a means to a better world for them, a world in which the questionable past (cheating) is easily forgotten as one's career advances?

As we shall see in the next section, there are exact counterparts in the scientific and engineering communities to the types of dishonesty exhibited by students, whether in science, engineering, or the humanities. Furthermore, it is simply not true that the sorts of pressures that invite cheating as a student are absent in the workplace. Smoothing data points on the graph of a freshman physics laboratory report to get an A on the report, selecting the research data that support the desired conclusion, entirely inventing the data, and plagiarism of the words and ideas of others all can advance one's academic aims, as can their obvious parallels in nonacademic settings.

2 Dishonesty in Research and Testing

Dishonesty in science and engineering takes several forms. Falsification of data, fabrication of data, plagiarism, and inappropriate attributions of authorship are among the most common, as shown in following.

- Falsification of data: distortion in representing data (e.g., omission of relevant data).
- Fabrication of data: making up data (e.g., for tests that were not actually done).
- Failure to respect intellectual property of others (e.g., violating trade secrets, patents, trademarks, and copyrights).
- Plagiarism: falsely appropriating the work of others as one's own.
- Inappropriate attribution of authorship (e.g., listing someone as an author who didn't warrant credit for this).

Falsification of data involves distorting data by smoothing out irregularities or presenting only those data that fit one's favored theory and discarding the rest. Whether such falsification has occurred is sometimes a matter of great controversy. Physicist Robert A. Millikan's famous oil drop experiment in the early part of the twentieth century is a case in point. This experiment was credited with showing the uniformity of the charge of electrons and Millikan received a Nobel Prize for his work. Complicating matters is a subsequent discovery that, despite claiming in his published report of the data that his findings were based on all the trials he conducted, his laboratory notebooks indicate that he did not include data on 49 of the 189 trials he conducted.

Defenders of Millikan point out that sometimes it is appropriate to leave out certain data, for example, when it is determined that the equipment used is not functioning as it should. Critics reply that if Millikan had such problems, he should have included this explanation of the exclusion of some of the data in his report. Further experimentation did confirm his conclusion about the uniformity of the charge of electrons. But some still dispute whether Millikan's own experiment should be credited with showing this.

Fabrication of data involves inventing data and even reporting results of experiments that were never conducted. The fabrication of data by psychologist Stephen Breuning illustrates the possibly far reaching harm that can result from this. In December 1983, Dr. Robert Sprague of the University of Illinois wrote an 8-page letter, with 44 pages of appendices, to the National Institute of Mental Health documenting the fraudulent work submitted to him by Breuning. Breuning claimed he was reporting on research he had conducted in a Coldwater, Michigan, mental facility on the effects psychotropic medication has on children in need of help. However, Breuning simply made up the data.

Even though Breunings admitted to fabricating his data only three months after Sprague sent his letter to NIMH, the case was not finally resolved until July 1989. During that 5½ years interval, Sprague himself was the first target of investigation, he had his own funded research severely curtailed, he was subjected to threats of lawsuits, and he had to testify before a U.S. House of Representatives committee. Most painful of all, Sprague's wife died in 1986, after a lengthy bout with diabetes. In fact, Sprague said, his wife's serious illness was one of the major factors prompting his whistle-blowing to NIMH. Realizing how dependent his diabetic wife was on reliable research and medication, Sprague was particularly sensitive to the dependency that children, and vulnerable populations in general, have on the trustworthiness of not only their caregivers but also on those who use them in experimental drug research.

Writing some years after the closing of the Breuning case, Sprague concluded his reflections on his own experiences by bringing to his readers attention other possible victims of Breuning's research misconduct namely, other psychologists and researchers who had collaborated with Breuning without being aware that he had fabricated data.

Psychologist Alan Poling, who at one time had Breuning as a student, has written about the consequences of Breuning's misconduct for his collaborators in research.

Strikingly, Poling points out that between 1979 and 1981, Breuning was a contributor to 34 percent of all published research on the psychopharmacological areas in which he was allegedly working. Of course, it does not follow that all of Breuning's publications are based on fabricated data, but determining which are and which are not is a time-consuming, demanding task. So, those who cited Breuning's publications in their own work may have wondered about the reliability of those publications that were not yet validated and they may have suffered from accusations of guilt by association. As Poling points out, this is especially unfair in those instances in which Breuning's collaborations with others involved no fraud at all. In short, Breuning's fabrications presented serious ethical issues from both utilitarian and respect for persons perspectives. For Breuning himself, shortcomings from a virtue perspective are evident.

Selected from Chapter 5 of "*Engineering Ethics: Concepts and Cases* (sixth edition)", C. E. Harris, M. S. Pritchard, R. W. James, *etc*, Nelson Education (2017).

Appendix Ⅰ. Best Schools in Chemical Engineering（化学工程高校排名）

(1) US News化学工程研究生院（部）排名

https://www.usnews.com/best-graduate-schools/top-engineering-schools/chemical-engineering-rankings

(2) Times Higher Education世界大学排名

https://www.timeshighereducation.com/world-university-rankings/by-subject

(3) QS Top Universities QS世界大学2023年化学工程学科排名

https://www.topuniversities.com/university-rankings/university-subject-rankings/2023/chemical-engineering

(4) US News全球化学工程与工艺专业大学排名

https://www.usnews.com/education/best-global-universities/chemical-engineering

(5) 全国化学工程与工艺专业大学排名（大学必备网）

https://www.dxsbb.com/news/7638.html

(6) 软科中国最好学科排名：化学工程与技术

https://www.shanghairanking.cn/rankings/bcsr/2022/0817

Appendix Ⅱ. How to Make Scientific Posters
（学术海报制作小技巧）

I Care About Your Audience

- If you are in a field that crosses disciplines you might need to write the paper a bit differently than if you were aiming the information at one field.
- In some cases, you may want your study to be accessible to all readers, so you need to write the paper in a way that everyone will understand.
- In other cases, the audience may be people in the same field that have background knowledge of the subject.
- Define all acronyms the first time you use the full word or phrase, then use the acronym throughout the rest of the paper.

JIN RESEARCH GROUP
INNOVATIVE CHEMISTRY & CATALYSIS

I.1 Be Active, NOT Positive: Use Active Voice

- For the most part, journals will require you to use active voice.
- This is not the case for every journal so make sure to check the style guide before you start writing.
- To use active voice, write statements like "We performed this experiment..." instead of "The experiment was performed".

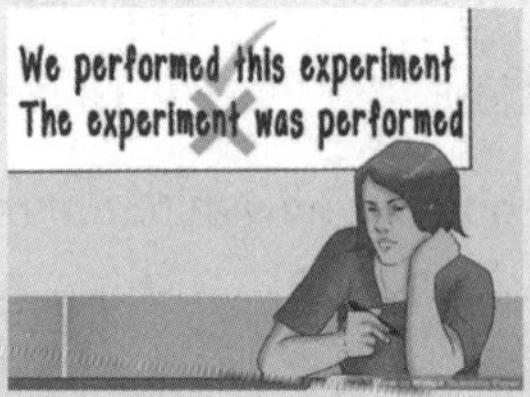

JIN RESEARCH GROUP
INNOVATIVE CHEMISTRY & CATALYSIS

I.2 Before You Get Started (Please Take Notes)

- Think as an Outsider --- Assume your audience know nothing
- Your audience care more about your motivation
- What are your key messages --- highlights in your research
- Some art work experience --- color code
- Draft a scheme --- the logic flow

JIN RESEARCH GROUP
INNOVATIVE CHEMISTRY & CATALYSIS

1.3 Follow the Rules, Please

- If you are submitting the paper to a journal for publication, they will have a style guide or author guide that details all of the necessary formatting.
- Maximum word count, order, the margin size, the font size/style, references format, etc.
- Following the author guide is absolutely necessary when submitting the paper for publication.
- Any restrictions on table/figure sizes or table/figure legends will also be included in the style guide.

JIN RESEARCH GROUP
INNOVATIVE CHEMISTRY & CATALYSIS

2 Writing Sections

- Start with the Materials and Methods section.
- Describe the results in the Results section.
- Interpret your data in the Discussion section.
- Review the literature in the Introduction.
- Summarize the paper in the Abstract.
- Write a descriptive Title.

JIN RESEARCH GROUP
INNOVATIVE CHEMISTRY & CATALYSIS

2.1 Materials and Methods: A Good Way to Make Your Boss Happy

- It is the easiest place to start and doesn't require much time to complete.
- The methods should be described clearly and anyone with the proper training should be able to replicate your experiments based on this section.
- Materials used for each method should be included, with references.
- Description of all statistical methods.

JIN RESEARCH GROUP
INNOVATIVE CHEMISTRY & CATALYSIS

2.2 Describe the Results

- Results should be reported neutrally and reference any tables and figures included in the paper.
- You don't need to include every experiment you performed or result you observed, just the information needed to convince the audience of your findings (Think critically & ahead).

JIN RESEARCH GROUP
INNOVATIVE CHEMISTRY & CATALYSIS

2.3 Interpret Your Data (Figures & Tables)

- The goal is to convince the reader that this data is important and they should care about it.
- Discuss them in the context of what is already known about your topic.
- Don't ignore other papers that contradict your findings.
- Draw quick conclusions about the data and talk about future experiments you would like to do to further the study.

JIN RESEARCH GROUP
INNOVATIVE CHEMISTRY & CATALYSIS

2.4 Review the Literature in the Introduction

- Three steps: your work is important and tell audience why it is important.
- General background & significance.
- The gap & issues & problems.
- State your hypothesis and objectives with highlights.

JIN RESEARCH GROUP
INNOVATIVE CHEMISTRY & CATALYSIS

2.5 Write a Descriptive Title

- Avoid technical jargon and abbreviations or acronyms.
- Consider which keywords you want to tag so that readers searching for articles on a specific topic will be directed to your work.

JIN RESEARCH GROUP
INNOVATIVE CHEMISTRY & CATALYSIS

3 Making the Figures and Tables

- Choose to present the data as a figure or a table.
- Format the table properly.
- Make data sets easily distinguishable.
- Include scale markers on photographs.
- Straightforward, NOT fluffy color code.
- Use fonts that are large enough to read.
- Write figure legends that describe the data.

JIN RESEARCH GROUP
INNOVATIVE CHEMISTRY & CATALYSIS

3.1 Choose to Present the Data

- Tables are used to present the raw data of the experiment while figures are used to illustrate comparisons/trend.
- Tables are frequently used to provide information about the makeup of a study group.
- Figures are used to compare the experimental results of the different groups in a visual way.

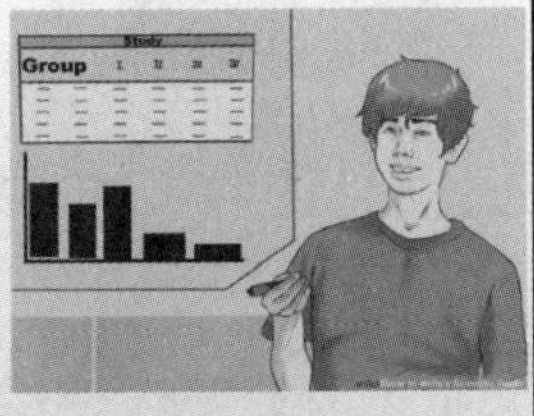

JIN RESEARCH GROUP
INNOVATIVE CHEMISTRY & CATALYSIS

3.2 Make Data Sets Easily Distinguishable

- When constructing a figure, avoid adding too many datasets to one graph.
- You may need to split the data into multiple graphs. Avoid adding more than 3~4 datasets per graph.
- Label all axes properly and use the appropriate scales.
- **Right & Comforting & Clear color code.**
- **Include (Scale) Markers on Photographs.**

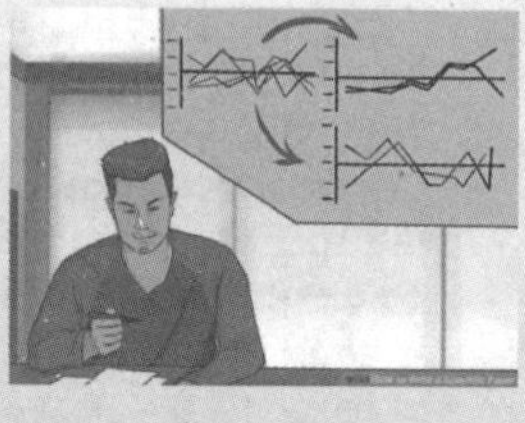

JIN RESEARCH GROUP
INNOVATIVE CHEMISTRY & CATALYSIS

3.3 Include Clear Fonts & Legends

- Use fonts that are large enough to read.
- When making the figure, the font may look perfectly legible.
- The figure legend shouldn't be too long.
- But it should give enough information to the reader for them to interpret the data without reading the corresponding text in the paper.

JIN RESEARCH GROUP
INNOVATIVE CHEMISTRY & CATALYSIS

4 Cite Your Sources Properly

- Use inline citations.
- Check the style guide for format.
- Match the content to the source.
- Use citation programs.

JIN RESEARCH GROUP
INNOVATIVE CHEMISTRY & CATALYSIS

Chapter 3 Solar Energy and Conversion Approaches（太阳能的转化利用）

California law mandates solar panels for new homes

Find more information at HuaNeng, a public wechat account.

Part A Comprehensive Reading

Solar Energy: A Brief Introduction

1 Background

The amount of solar radiant energy falling on a surface per unit area and per unit time is called irradiance. The mean extraterrestrial irradiance normal to the solar **beam** on the outer fringes of the earth's atmosphere is approximately 1.36kW/m^2. Since the earth's orbit is elliptical, the sun-to-earth distance varies slightly with time of year, and the actual extraterrestrial irradiance varies by ±3.4% during the year. The angle subtended by the sun when viewed from the earth is only 0.0093rad (approximately 32min of arc); the direct beam radiation reaching the earth is therefore almost parallel. Although the brightness of the solar disc decreases from center to edge, for most engineering calculations, the disc can be assumed to be of uniform brightness. The radiant energy from the sun is distributed over a range of wavelengths, and the energy falling on a unit surface area per unit time within a particular spectral band is known as the spectral irradiance; its value is usually expressed in watts per square meter per nanometer of **bandwidth**. The extraterrestrial spectral irradiance is shown in Figure 1. It can be approximated by the **spectrum** of a black body at 5800K. In the upper part of Figure 1, the wave bands typically usable for different solar applications are shown, and the lowest curve shows the spectral direct-beam irradiance at sea level on earth under clear sky conditions with the sun overhead.

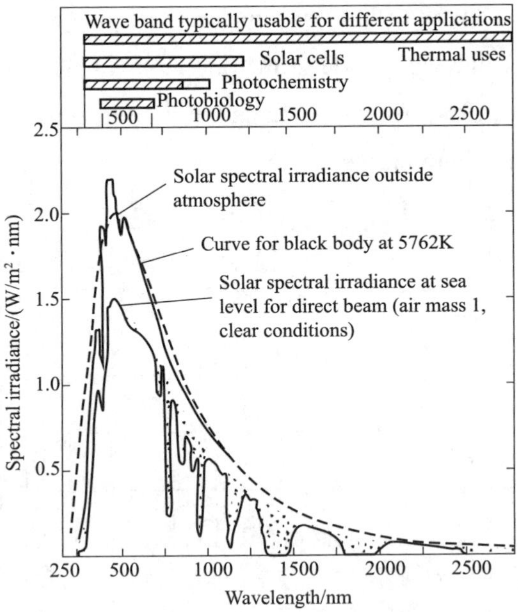

Figure 1 Spectral irradiance curves for direct sunlight are extraterrestrial and at sea level with the sun directly overhead. Shaded areas indicate absorption owing to atmospheric constituents, mainly H_2O, CO_2, and O_3. Wavelengths potentially utilized in different solar energy applications are indicated at the top

The amount of sunlight striking the earth's atmosphere continuously is 1.75×10^5TW. Considering a 60% **transmittance** through the atmospheric cloud cover, 1.05×10^5TW reaches the earth's surface continuously. If the irradiance on only 1% of the earth's surface could be converted into electric energy with a 10% efficiency, it would provide a resource base of 10^5TW, while the total global energy needs for 2050 are projected to be approximately 25~30TW. The present state of solar energy technologies is such that commercial solar panel efficiencies have reached more than 20%, laboratory multijunction solar cell efficiencies under concentrating sun have exceeded 40%, and solar thermal systems provide efficiencies of 40%~60%. Solar **photovoltaic panels** have come down in cost from approximately \$30/W to approximately \$0.50/W in the last three decades. At \$0.50/W panel cost, the overall system cost is around \$2/W, which is already lower than **grid** electricity in the Caribbean island communities. Of course, there are many off-grid applications where solar photovoltaic is already **cost–effective**. With net metering and governmental incentives, such as feed-in laws and other policies, grid-connected applications such as

Building-Integrated photovoltaic have become cost-effective even where grid electricity is cheaper. As a result, the worldwide growth in photovoltaic production has averaged more than 43% per year from 2000 to 2012 and 61% from 2007 to 2012 (Figure 2) with Europe showing the maximum growth.

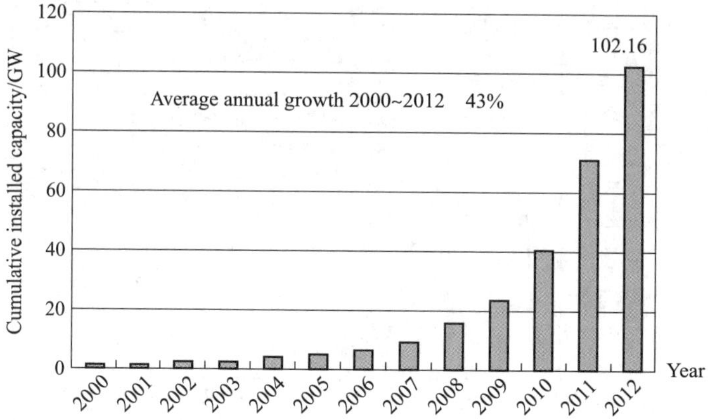

Figure 2 World solar photovoltaic production 2000~2012 (GWp) (From EPIA, Global Market Outlook for Photovoltaics, European Photovoltaic Industries Association, Brussels, Belgium, http://www.epia.org.)

Solar thermal power using concentrating solar collectors was the first solar technology that demonstrated its grid power potential. A 354MW_e concentrating solar thermal power (CSP) plant has been operating continuously in California since 1988. Progress in solar thermal power stalled after that time because of poor policy and lack of research and development (R&D).

However, the last 10 years have seen a resurgence of interest in this area and a number of solar thermal power plants around the world are under construction. The largest CSP plant with a capacity of 400MW came on line in Nevada in February 2014. The cost of power from these plants (which is so far in the range of 12 to 16 US cents/$kW{\cdot}h_e$) has the potential to go down to 5 US cents/$kW{\cdot}h_e$ with scale-up and creation of a mass market. An advantage of solar thermal power is that thermal energy can be stored efficiently, and fuels such as natural gas or biogas may be used as backup to ensure continuous operation. If this technology is combined with power plants operating on fossil fuels, it has the potential to extend the time frame of the existing fossil fuels. Low-temperature solar thermal systems and applications have been well developed for quite some time. They are being actively installed wherever the policies favor their deployment. Figure 3 gives an idea of the rate of growth of solar thermal systems in the world. We shall now describe, briefly, the important conversion methods.

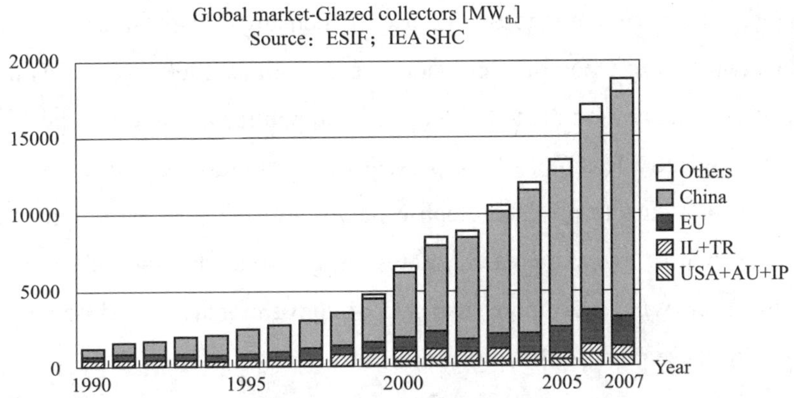

Figure 3 Deployment of solar heat (glazed) collectors, /MW_{th}

2 Thermal Conversion

Thermal conversion is an engineering conversion method that utilizes a familiar scientific phenomenon. When a dark surface is placed in sunshine, it absorbs solar energy and heats up. Solar energy collectors working on this principle consist of a surface facing the sun, which transfers part of the energy it absorbs to a working fluid in contact with it. To reduce heat losses to the atmosphere, one or two sheets of glass are usually placed over the absorber surface to improve its efficiency. These types of thermal collectors suffer from heat losses owing to radiation and convection, which increase rapidly as the temperature of the working fluid increases. Improvements such as the use of selective surfaces, evacuation of the collector to reduce heat losses, and special kinds of glass are used to increase the efficiency of these devices.

The simple thermal conversion devices described above are called flat plate collectors. They are available today for operation over a range of temperatures up to approximately 365K (200°F). These collectors are suitable mainly for providing hot water service and space heating and possibly are also able to operate absorption-type air-conditioning systems.

The thermal utilization of solar radiation for the purpose of generating low temperature heat is at the present time technically feasible and economically viable for producing hot water and heating swimming pools. In some parts of the world, thermal low-temperature utilization is also used for heating and cooling buildings.

The generation of higher working temperatures as needed, for example, to operate a conventional steam engine requires the use of focusing devices in connection with a basic absorber-receiver. Operating temperatures as high as 4000K (6740°F) have been achieved, and the generation of steam to operate pumps for irrigation purposes has also proved technologically feasible. At the present time, a number of focusing devices for the generation of steam to produce electric power are under construction in different regions of the world, and the cost estimates suggest that the cost of solar power in favorable locations will be no more than that of conventional thermal power when the development of these plants has been completed.

3 Photovoltaic Conversion

The conversion of solar radiation into electrical energy by means of solar cells was developed as a part of satellite and space-travel technology. The present technology utilizes the band gap of semiconductor materials to generate electrons and conduct them in a preferred direction using a **p-n junction** (described in detail later in the book). The theoretical maximum efficiency of single solar cells is approximately 33%, and in practice, efficiencies as high as 25% have been achieved with silicon photovoltaic devices. Multijunction solar cells can utilize a much broader spectrum of solar energy at a higher efficiency. Efficiencies as high as 44% have been achieved in the laboratory and the present research is aimed at efficiencies of around 50%. The overall system efficiencies of the commercially available photovoltaic panels today are in the range of 10% to 20%. With the development of thin film solar photovoltaic technologies, the costs of panels have come down from approximately \$30/W_p to as low as \$0.50/W_p. Here, W_p stands for peak watt, which means that the stated output would be expected when the solar irradiance on the panel is 1000W/m^2 at standard atmospheric conditions. This level of solar radiation would be expected at a sunny location on a clear day around noon while the panel is directly facing the sun. With costs of the panels being lower than \$1/W_p, the total photovoltaic system costs are in the range of \$1.50/W_p (for multi-megawatt scale systems) to approximately \$3/W_p for residential-scale systems.

Because of the reduction in costs, improvements in reliability and lifetimes, and preferential public policies, the photovoltaic industry is growing at an average annual rate of more than 50%.

Another way of direct conversion of sunlight to electricity takes advantage of the electromagnetic wave nature of sunlight as opposed to the quantum nature used in the conventional photovoltaic. This concept proposes to use nanoscale and microscale antennas (size of the order of the wavelength of radiation) to convert solar and ambient radiation into electricity. The proposed panel would consist of billions of such antennas coupled with rectifiers. The combined antenna and rectifier structure is known as a rectenna. The theoretical efficiency of a rectenna is in the range of 80% to 90%. An efficiency of greater than 80% was demonstrated for a rectenna converting radiation at a frequency of 2.4GHz to electricity. However, since the frequency of visible solar radiation is in the range of 300THz, which would require a five-orders-of-magnitude size reduction of antennas and a totally different concept of the rectifier, development of this concept for converting solar energy is no simple task and is expected to take probably a decade or longer to become practical.

4 Limitations of Solar Energy

The first problem encountered in the engineering design of equipment for solar energy utilization is the low flux density, which makes necessary large surfaces to collect solar energy for large-scale utilization. Also, the larger the surfaces, the more expensive the delivered energy becomes. When the sun is directly overhead on a cloudless day, $10m^2$ of surface could theoretically provide energy at a 70% efficiency of collection and 30% conversion efficiency at a rate of approximately 2kW. Several factors reduce this amount in practice, however.

The total solar energy reaching the earth is made up of two parts: the energy in the direct beam and the diffuse energy from the sky. The amount of direct energy depends on the cloudiness and the position of the sun and is obviously greatest on clear days. Some solar radiation falling on clouds is diffused by scattering, but clouds do not absorb all of the energy. The effect of clouds is mainly to increase the percentage of diffuse energy in the total energy reaching the surface, and diffuse irradiance in summer months with high and broken clouds can be as high as 400W/m^2. Thick clouds let less energy pass than thin clouds and scatter proportionally more of the total energy back into space.

The second practical limitation that is not apparent from the macroscopic energy view is that most of the solar energy falls on remote areas and would therefore require some means of transmission to be useful to the industrialized nations. The mean amount of energy available on a horizontal plane is greatest in the continental desert areas around latitudes 25°N and 25°S of the equator and falls off toward both the equator and the poles. The highest annual mean irradiance is 300W/m^2 in the Red Sea area. Clouds reduce the mean global irradiance considerably in **equatorial** regions. However, whereas in northern climates the mean horizontal surface global irradiance varies from season to season, it remains relatively constant in equatorial regions. Typical values of mean annual horizontal surface irradiance are as follows: Australia, approximately 200W/m^2; United States, 185W/m^2; United Kingdom, 105W/m^2.

Table 1 shows the total solar electric energy potential for the solar energy-rich areas of the world. It is important to note that many of these parts of the world, for example, Saudi Arabia, central Australia, and parts of South Africa and India, are flat deserts that are practically unusable for agriculture. At the same time, it should be noted that these are also regions with little or no water, which point to the need to develop dry cooling options for solar thermal electric power. Although the global distribution of solar energy does not favor the industrialized parts of the world, it is important to note that these are the regions, especially Europe, that have led in the deployment of solar power over the last decade.

Table 1. Solar-electric energy from Earth's high-insolation area

Desert	Nominal areas / km^2	Nominal annual thermal energy flux / [GW·h(th)/km^2]	Percentage of area assumed useable	Electric energy extracted at 25% efficiency/ (GW·h/year)
North Africa	7770000	2300	15	670000000
Arabian Peninsula	1300000	2500	30	244000000
Western and central Australia	1550000	2000	25	194000000
Kalahari	518000	2000	50	129000000
Thar	259000	2000	50	65000000

(continued)

Desert	Nominal areas / km^2	Nominal annual thermal energy flux / [GW·h(th)/km^2]	Percentage of area assumed useable	Electric energy extracted at 25% efficiency/(GW·h/year)
Mojave, southern California	35000	2200	20	3900000
Vizcaino, Baja, California	15500	2200	25	2100000
Total/Average	11447500	2190	31	1303000000

a: Parts of Arabian and Libyan deserts.

b: Approximately 60% of Rub'al Khali desert.

The third limitation of solar energy as a large-scale source of power and heat is its intermittency. Solar energy has a regular daily cycle owing to the turning of the earth around its axis and a regular annual cycle owing to the inclination of the earth axis with the plane of the ecliptic and to the motion of the earth around the sun, and is also unavailable during periods of bad weather. It is obvious that these daily and seasonal variations in irradiance, exacerbated by variations owing to weather, introduce the need for storage or a supplementary fuel.

Selected from Chapter 8 of "*Principles of Solar Engineering* (third edition)", D. Yogi Goswami, CRC Press Taylor & Francis Group (2014).

Notes

Beam: *noun.* (1) a long, sturdy piece of squared timber or metal spanning an opening or part of a building, usually to support the roof or floor above. (2) a ray or shaft of light. (3) a radiant or good-natured look or smile. *verb.* (1) transmit (a radio signal or broadcast) in a specified direction. (2) (of a light or light source) shine brightly. (3) smile radiantly.

Bandwidth: *noun.* (1) a range of frequencies within a given band, in particular that used for transmitting a signal. (2) the energy or mental capacity required to deal with a situation.

Spectrum: *noun.* (1) a band of colors, as seen in a rainbow, produced by separation of the components of light by their different degrees of refraction according to wavelength. (2) used to classify something, or suggest that it can be classified, in terms of its position on a scale between two extreme or opposite points.

Transmittance: *noun.* the ratio of the light energy falling on a body to that transmitted through it.

Photovoltaic panel: convert the sun's energy into electrical energy

Grid: *noun.* (1) a framework of spaced bars that are parallel to or cross each other; a grating. (2) a network of lines that cross each other to form a series of squares or rectangles. (3) a network of cables or pipes for distributing power, especially high-voltage transmission lines for electricity. *adjective.* put into or set out as a grid.

Cost-effective: *adjective.* effective or productive in relation to its cost.

Equatorial: *adjective.* of, at, or near the equator.

Questions

(1) Why does solar energy be viewed as renewable energy?

(2) What is the range of wavelength for the radiant energy from the sun?

(3) Considering the amount of sunlight striking the earth's atmosphere continuously is 1.75×10^5 TW, what is the amount, which could be converted into electric energy, on the surface of earth?

(4) What are the advantages and limitations of solar thermal power?

(5) Thermal conversion of thermal collectors is one of the most mature ways to take use of solar energy. However, it suffers from energy loss. Please analyze the reasons of energy loss and the methods to improve the performance of the thermal collectors.

(6) What is the basic theory of photovoltaic conversion of solar cells?

(7) What are the advantages of film solar photovoltaic technologies?

(8) Although there are plenty of advantages of solar energy, there are still lots of limitations of solar energy utilization. Please state the limitations and provide some examples.

Part B Reading for Innovation

Solar Thermal Power

1 Thermodynamic Power Cycles

There are two basic approaches to solar electric power generation. One is by photovoltaic process, a direct energy conversion, which will be described in detail in this chapter. The other approach is to convert sunlight to heat at high temperatures and then heat to mechanical energy by a thermodynamic power cycle and, finally, convert the **mechanical energy** to electricity. This indirect approach, called solar thermal power or CSP, is based on well-established principles of thermal power. A vast majority of electricity in the world is produced by thermal power conversion. Most of the thermal power production in the world is based on Rankine cycle and, to a smaller extent, Brayton cycle. Both of these are applicable to solar thermal power conversion, with Rankine cycle being the most popular. Normally, water is the working fluid for the Rankine cycle. However, for lower-temperature solar collection systems (70°C to approximately 300°C). Organic fluids are used, in which case the cycle is commonly referred to as the organic Rankine Cycle (ORC). When a Rankine cycle is operated under supercritical conditions of the working fluid, the cycle is usually referred to as the Supercritical Rankine Cycle (SRC). The Stirling cycle has also shown great potential, and solar thermal power systems based on this cycle are under development. More recently, researchers have been developing modifications of these cycles or entirely new cycles or combined cycles to increase the conversion efficiencies and to make them more applicable to the solar collection systems.

(1) Rankine Cycle

Most of the existing thermal power plants are based on the Rankine cycle. The basic ideal Rankine cycle is shown in Figure 1, which also shows a Temperature-entropy (T-s) diagram for steam as a working fluid. The ideal cycle consists of the following process:

1~2 Saturated liquid from the condenser at state 1 is pumped to the boiler at state 2 isentropically.

2~3 Liquid is heated in the boiler at constant pressure. The temperature of the liquid rises until it becomes a saturated liquid. Further addition of heat vaporizes the liquid at constant temperature until all of the liquid turns into saturated vapor. Any additional heat superheats the working fluid to state 3.

3~4 Steam expands isentropically through a turbine to state 4.

4~1 Steam exiting the turbine is condensed at constant pressure until it returns to state 1 as saturated liquid.

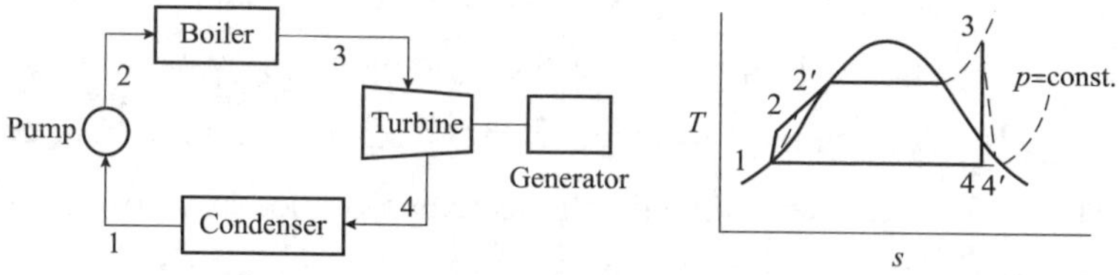

Figure 1　Basic Rankine power cycle

(2) Components of a Rankine Power Plant

Major components of a Rankine power plant include boiler, **turbine**, condenser, pumps (condensate pump, feedwater booster, and boiler feed pump), and heat exchangers (open heaters and closed heaters). All of the components of a solar thermal power plant are the same as those in a conventional thermal power plant except the boiler. The boiler in a solar thermal power plant includes a solar collection system, a storage system, an auxiliary fuel heater and heat exchangers. Figure 2 shows a schematic representation of a solar boiler. The maximum temperature from the solar system depends on the type of solar collection system (PTCs, central receiver with heliostat field, parabolic dishes, etc.). A fossil fuel or a biomass fuel may be used as the auxiliary fuel. Figure 3 shows the types of condenser or heat rejection used in Rankine cycle solar power systems.

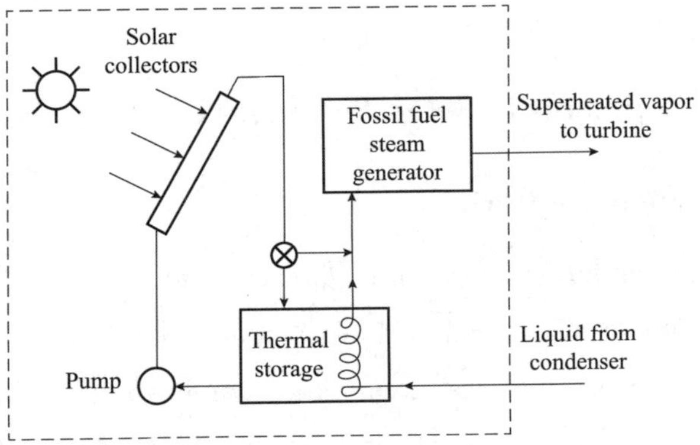

Figure 2 Schematic of a solar boiler

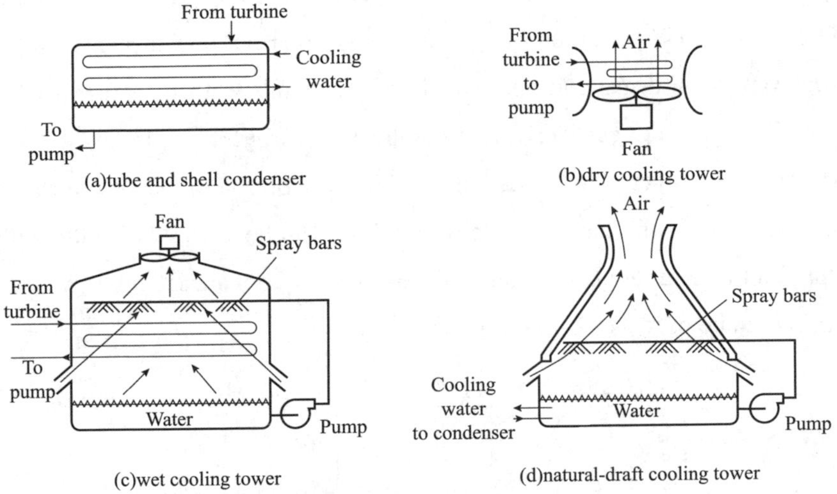

Figure 3 Types of condenser or heat rejection used in Rankine cycle solar power systems(Stine, W.B. and R.W. Harrigan: *Solar Energy Fundamentals and Design: With Computer Applications*. 1985. Copyright Wiley–VCH Verlag GmbH & Co. KGaA.)

A condenser is a large heat exchanger that condenses the exhaust vapor from the turbine. Steam turbines employ surface-type condensers, mainly shell and tube heat exchangers operating under vacuum. The vacuum in the condenser reduces the exhaust pressure at the turbine blade exit to maximize the work in the turbine. The cooling water from either a large body of water such as a river or a lake, or from cooling tower, circulates through the condenser tubes. The cooling water is cooled in the cooling tower by evaporation. The air flow in the cooling tower is either natural draft (hyperbolic towers) or forced draft. The condensate and feedwater pumps are motor-driven centrifugal pumps, while the boiler feed pumps may be motor- or turbine-driven centrifugal pumps.

2 Recent Developments in Solar Thermal Power Cycles

(1) Hybrid and Combined Cycles

If the supplemental natural gas flows in the power system are large enough, it makes sense to use the natural gas in a combined-cycle mode. With the present-day technology of natural gas combustion turbines, a combined combustion turbine/steam Rankine cycle power plant can achieve a cycle efficiency as high as 58%. Figure 4 shows a schematic of a combined-cycle solar/natural gas power plant. In 2012, Florida Power & Light built a 75MW parabolic trough-based solar thermal power plant as a preheater for a 1000MW natural gas combined-cycle power plant. With the present cost of the commercially available, aeroderivative combustion turbines as low as \$500/kW, and the cost of a solar power plant being approximately \$2500~\$3000/kW, the capital costs of a combined cycle solar/natural gas power plant could be as low as \$1000~\$2000/kW, depending on the natural gas fraction. This idea seems very attractive, especially where natural gas is to be used as a backup fuel anyway.

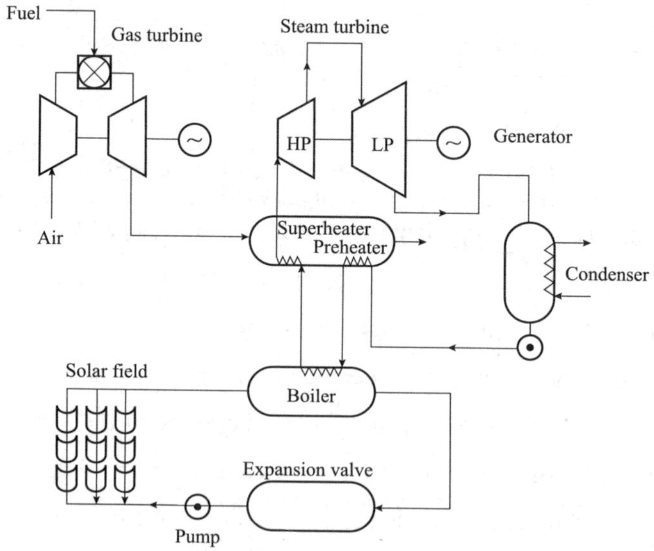

Figure 4 Concept of an integrated solar combination-cycle power system (From Washam, B.M. et al., *Integrated Solar Combined Cycle Systems [ISCCS] Utilizing Solar Parabolic Trough Technology-Golden Opportunities for the 90s*. Diablo, CA: Spencer Management Associates, 1993.)

(2) Kalina Cycle

Kalina (1984) proposed a novel bottoming cycle for a combined-cycle power plant that improves the overall performance of the cycle. This novel cycle, now known as the Kalina cycle, uses a mixture of ammonia and water as the working fluid for the bottoming cycle expansion turbine. Using a two-component working fluid and multipressure boiling, one can reduce heat transfer-related **irreversibilities** and therefore improve the resource effectiveness. Also, the ammonia-water mixture can be used as a working fluid for a lower-temperature (250°C) solar energy system employing lower-cost parabolic trough technology. According to Kalina and Tribus (1992), it is possible to improve the efficiency of the bottoming cycle by more than 45% over the steam Rankine cycle for a lower temperature resource (~250°C). Therefore, if the Kalina cycle is used for solar thermal power in place of the Rankine cycle, the capital cost could be reduced by at least 45% because of the higher efficiency and the possibility of using cheaper concentrating collectors owing to the lower temperature requirement. These improvements may be enough to make solar thermal power economically competitive by itself and not a burden on natural gas for combined-cycle power plants. This offers an opportunity for research in solar thermal power utilizing the Kalina cycle.Figure 5 shows heat transfer between a heat source and a working fluid.

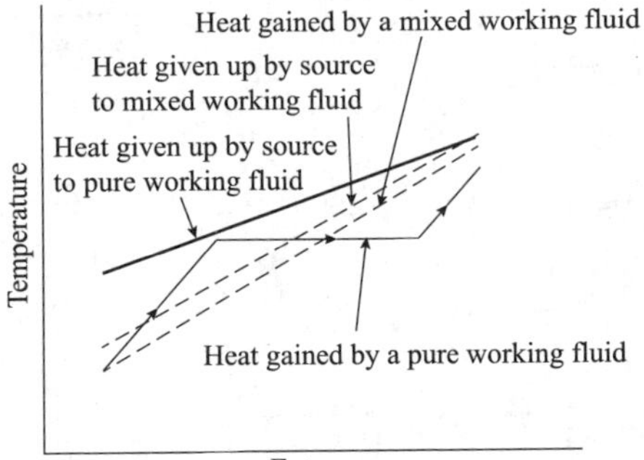

Figure 5 Heat transfer between a heat source and a working fluid

(3) Combined Power/Cooling Cycle

Goswami (1995) has proposed a new thermodynamic cycle that improves the cycle efficiency and, therefore, resource utilization by producing power and refrigeration in the same cycle. The new cycle, also known as the Goswami cycle, was developed to use ammonia as the working fluid in an innovative combination of an ammonia-based Rankine cycle and an ammonia-water absorption refrigeration cycle. The use of any mixture of fluids was later allowed provided that one is much more volatile than the other and the more volatile component could be condensed by absorption in the other. The ammonia-water mixture is an obvious example, but other examples include mixtures of hydrocarbons such as isobutane and heptane or pentane and decane. Figure 6 shows a schematic of the cycle. By boiling a mixture, for example, ammonia-water, this cycle reduces the irreversibilities associated with the heat transfer from a sensible heat source. Also, by expanding almost pure ammonia in the last-stage turbine, it allows the working fluid to go down to a much lower temperature, providing refrigeration. The working fluid is condensed by absorption in the water. The net effect lowers the sink temperature of the cycle. The cycle can use source temperatures as low as 100°C, thereby making it a useful power cycle for low-cost solar thermal collectors, geothermal resources, and waste heat from existing power plants.

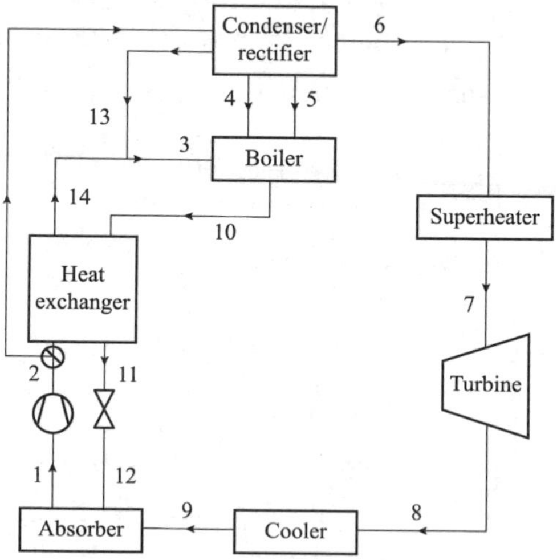

Figure 6　A schematic diagram of the Goswami combined power/cooling cycle

Selected from Chapter 8 of “*Principles of Solar Engineering* (third edition)”, D. Yogi Goswami, CRC Press Taylor & Francis Group (2014).

Important Words and Phrases

Mechanical energy: the sum of kinetic energy and potential energy in an object that is used to do a particular work. In other words, it describes the energy of an object because of its motion or position, or both.

Entropy: *noun.* (1) a thermodynamic quantity representing the unavailability of a system's thermal energy for conversion into mechanical work, often interpreted as the degree of disorder or randomness in the system. (2) lack of order or predictability; gradual decline into disorder. (3) (in information theory) a logarithmic measure of the rate of transfer of information in a particular message or language.

Turbine: *noun.* a machine for producing continuous power in which a wheel or rotor, typically fitted with vanes, is made to revolve by a fast-moving flow of water, steam, gas, air, or other fluid.

Irreversibility: *adjective.* not able to be undone or altered.

Questions

(1) What is the definition of Rankine cycle? What are the processes included in the ideal Rankine cycle?

(2) How many components of a Rankine power plant included? What are they?

(3) Please describe with your own words how a solar boiler work.

(4) What is the concept of an integrated solar combination-cycle power system?

(5) Do you know Kalina' s story? Please describe how Kalina cycle was developed and what Kalina cycle is about.

(6) What is Goswami cycle? What are the differences between Goswami cycle and Kalina cycle?

(7) The type of the working fluid has significant influence to the performance. What type of working fluid showed good performance for the Goswami combined power/cooling cycle?

Part C Reading for Fun

The Historical Perspective

Attempts to harness the sun's energy for power production date back to at least 1774 (Trombe 1956) when the French chemist Lavoisier and the English scientist Joseph Priestley discovered oxygen and developed the theory of combustion by concentrating the rays of the sun on mercuric oxide in a test tube, collecting the gas produced with the aid of solar energy, and burning a candle in the gas. Also, during the same year, an impressive picture of Lavoisier was published in which he stands on a platform near the focus of a large glass lens and is carrying out other experiments with focused sunlight [see schematic, Figure 1(a)].

A century later, in 1878, a small solar power plant was exhibited at the World's Fair in Paris [Figure 1(b)]. To drive this solar steam engine, sunlight was focused from a parabolic reflector onto a steam boiler located at its focus; this produced the steam that operated a small reciprocating steam engine that ran a printing press. In 1901, a 10hp solar steam engine was operated by A.G. Eneas in Pasadena, California (Daniels 1964). It used a $700ft^2$ focusing collector the shape of a truncated cone as shown in Figure 2. Between 1907 and 1913, the American engineer F. Shuman developed solar-driven hydraulic pumps; in 1913, he built, jointly with C.V. Boys, a 50hp solar engine for pumping irrigation water from the Nile near Cairo in Egypt (Figure 3). This device used long parabolic troughs that focused solar radiation onto a central pipe with a concentration ratio of 4.5 : 1.

With the increasing availability of low-cost oil and natural gas, interest in solar energy for power production waned. Except for C.G. Abbott, who exhibited in 1936 a $^1/_2$hp solar-powered engine at an International Power Conference in Washington,

D.C., and in 1938 in Florida, an improved, somewhat smaller version with a flash boiler, there was very little activity in the field of solar power between 1915 and 1950. Interest in solar power revived in 1949 when, at the centennial meeting of the American Association for the Advancement of Science in Washington, D.C., one session was devoted to future energy sources. At that time, the potentials as well as the economic problems of solar energy utilization were clearly presented by Daniels (1949).

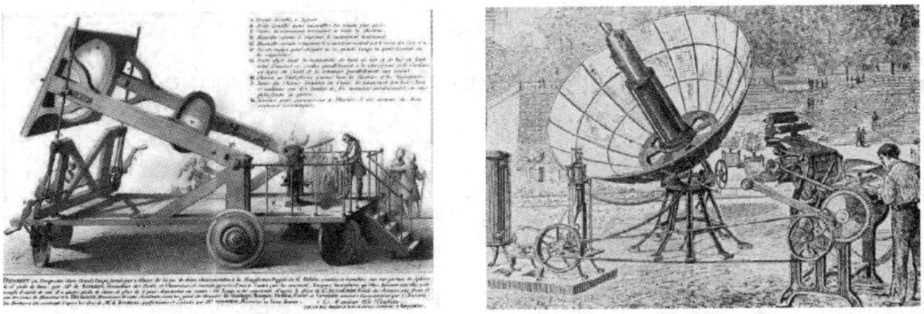

(a) (b)

Figure 1 (a) Solar furnace used by Lavoisier in 1774. (Courtesy of Bibliotheque Nationale de Paris. Lavoisier, Oeuvres, vol. 3.) (b) Parabolic collector powered a printing press at the 1878 Paris Exposition (From Abbot, C.G., *The Sun and the Welfare of Man*, Washington, DC: Smithsonian Institution, 1929.)

Figure 2 Irrigation pumps were run by a solar–powered steam engine in Arizona in the early 1900s. The system consisted of an inverted cone that focused rays of the sun on the boiler (From Abbot, C.G., *The Sun and the Welfare of Man*, Washington, DC: Smithsonian Institution, 1929.)

Some important conferences that considered solar power generation were held by UNESCO in 1954, the Association for Applied Solar Energy in 1955, the US National Academy of Sciences in 1961, and the United Nations in 1961. In addition, a research and development program supported by the National Aeronautics and Space Administration to build a solar electric power system capable of supplying electricity for the US space program was undertaken in the 1960s. However, widespread interest developed only after research funds became available for the development of earth-bound solar electric power and process heat after the oil embargo in 1973. Pace of development increased in the 1980s with the development of commercial parabolic trough collector (PTC)-based power plants SEGS I to SEGS IX totaling 354MW_e capacity and the 1MW_e Solar I central receiver tower demonstration plant in California. However, the progress stalled after that until a commercial 75MW_e parabolic trough power plant started operating near Las Vegas, Nevada, in 2007. Since then, commercial and research activity in concentrated solar thermal power, also known as CSP, has increased tremendously in Europe, the United States, Asia, and Australia.

Figure 3 Solar irrigation pump (50hp) operating in 1913 in Egypt (From Abbot, C.G., *The Sun and the Welfare of Man*, Washington, DC: Smithsonian Institution, 1929.)

Currently, four different CSP technologies are in use based on the type of solar collection system: (1) PTC, (2) linear Fresnel reflector (LFR) system, (3) power tower or central receiver system (CRS), and (4) parabolic dish system. Of these, the PTC-based plants are the most in operation since 1985. The LFR system is in principle the same as the PTC except that the parabolic concentration is achieved via LFRs. CRS systems have the potential to achieve the highest efficiencies, which is why most of the present research activity in CSP is focused on CRS-based systems.

Selected from Chapter 8 of “*Principles of Solar Engineering* (third edition)”, D. Yogi Goswami, CRC Press Taylor & Francis Group (2014).

Solar Cell: The Fundamentals

1 The Working Principle

The working principle of solar cells is based on the photovoltaic effect, i.e. the generation of a potential difference at the junction of two different materials in response to electromagnetic radiation. The photovoltaic effect is closely related to the photoelectric effect, where electrons are emitted from a material that has absorbed light with a frequency above a material-dependent threshold frequency. In 1905, Albert Einstein understood that this effect can be explained by assuming that the light consists of well-defined energy quanta, called photons. The energy of such a photon is given by:

$$E=h\nu$$

where h is Planck's constant and ν is the frequency of the light. For his explanation of the photoelectric effect Einstein received the Nobel Prize in Physics in 1921.

Absorption of a photon in a material means that its energy is used to excite an electron from an initial energy level E_i to a higher energy level E_f, as shown in Figure 1. Photons can only be absorbed if electron energy levels E_i and E_f are present so that their difference equals the photon energy, $h\nu=E_f-E_i$. In an ideal semiconductor, electrons can populate energy levels below the so-called valence band edge, E_V, and above the so-

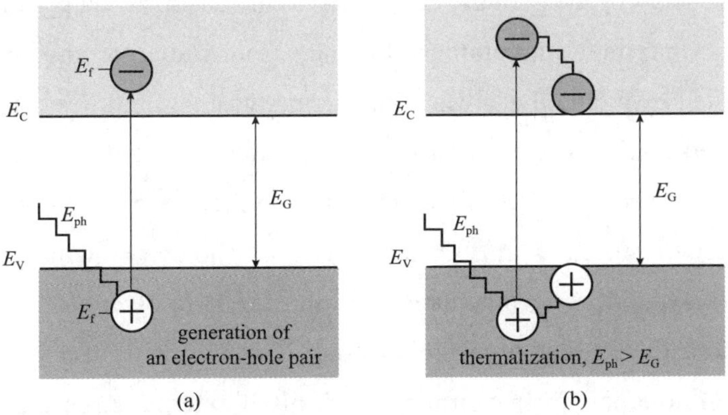

Figure 1 (a) Illustrating the absorption of a photon in a semiconductor with bandgap E_G. The photon with energy $E_{ph}=h\nu$ excites an electron from E_i to E_f. At E_i a hole is created. (b) If $E_{ph}>E_G$, a part of the energy is thermalized called conduction band edge, E_C. Between those two bands no allowed energy states exist which could be populated by electrons. Hence, this energy difference is called the bandgap, $E_G=E_C-E_V$. If a photon with an energy smaller than E_G reaches an ideal semiconductor, it will not be absorbed but will traverse the material without interaction.

In a real semiconductor, the valence and conduction bands are not flat, but vary depending on the so-called k-vector that describes the momentum of an electron in the semiconductor. This means that the energy of an electron is dependent on its momentum because of the periodic structure of the semiconductor crystal. If the maximum of the valence band and the minimum of the conduction band occur at the same k-vector, an electron can be excited from the valence to the conduction band without a change in the momentum. Such a semiconductor is called a direct bandgap material. If the electron cannot be excited without changing its momentum, we refer to it as an indirect bandgap material. The electron can only change its momentum by momentum exchange with the crystal, i.e. by receiving momentum from or giving momentum to vibrations of the crystal lattice. The absorption coefficient in a direct bandgap material is much higher than in an indirect bandgap material, thus the absorbing semiconductor, often just called the absorber, can be much thinner.

If an electron is excited from E_i to E_f, a void is created at E_i. This void behaves like a particle with a positive elementary charge and is called a hole. The absorption of a photon therefore leads to the creation of an electron-hole pair, as illustrated in Figure 2. The radiative energy of the photon is converted to the chemical energy of the electron-hole pair. The maximal conversion efficiency from radiative energy to chemical energy is limited by thermodynamics. This thermodynamic limit lies between 67% for nonconcentrated sunlight and 86% for fully concentrated sunlight.

Usually, the electron-hole pair will recombine, i.e. the electron will fall back to the initial energy level E_i, as illustrated in Figure 2. The energy will then be released either as photon (radiative recombination) or transferred to other electrons or holes or lattice vibrations (non-radiative recombination). If one wants to use the energy stored in the electron-hole pair for performing work in an external circuit, semipermeable membranes must be present on both sides of the absorber, such that electrons can only flow out through one membrane and holes can only flow out through the other membrane, as illustrated in Figure 2. In most solar cells, these membranes are formed by *n*- and *p* type materials.

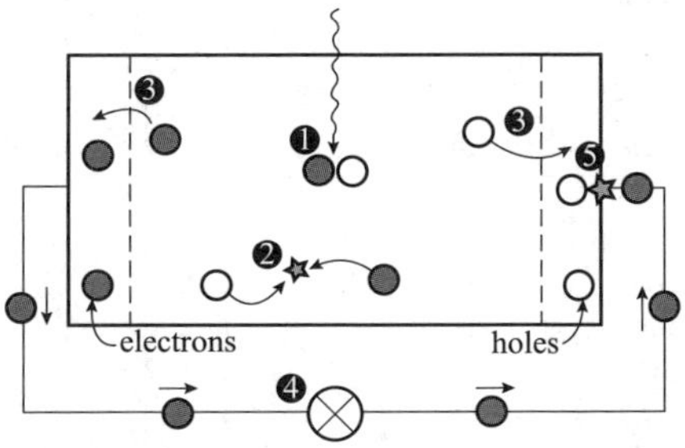

Figure 2 A very simple solar cell model. Absorption of a photon leads to the generation of an electron–hole pair. Usually, the electrons and holes will recombine. With semipermeable membranes the electrons and the holes can be separated. The separated electrons can be used to drive an electric circuit. After the electrons have passed through the circuit, they will recombine with holes.

Selected from "*Solar Energy: The physics and Engineering of Photovoltaic Conversion Technologies and Systems*". Arno HM Smets, et al. UIT Cambridge England.

2 Semiconductors

For any atom, the electrons arrange themselves in orbitals around the nucleus so as to result in the minimum amount of energy. Figure 3 shows the distribution of the electrons in various shells and subshells in light elements. In elements that have electrons in multiple shells, the innermost electrons have minimum energy and, therefore, require the maximum amount of externally imparted energy to overcome the attraction of the nucleus and become free. Electrons in the outermost band of subshells are the only ones that participate in the interaction of an atom with its neighboring atoms. If these electrons are very loosely attached to the atom, they may attach themselves with a neighboring atom to give that atom a negative charge, leaving the original atom as a positive charged ion. The positive and negatively charged ions become attached by the force of attraction of the charges, thus forming ionic bonds. If the electrons in the outermost band do not fill the band completely but are not loosely attached either, they arrange themselves so that neighboring atoms can share them to make the outermost bands full. The bonds thus formed between the neighboring atoms are called covalent bonds.

Principal quantum number (n)				1	2		3		4		5
Azimuthal quantum number (l)				0	0	1	0	1	2	0	1
Letter designation of state				1s	2s	2p	3s	3p	3d	4s	4p
Z	Symbol	Element	V_i Volts								
1	H	Hydrogen	13.60	1							
2	He	Helium	24.58	2							
3	Li	Lithium	5.39		1						
4	Be	Beryllium	9.32		2						
5	B	Boron	8.30		2	1					
6	C	Carbon	11.26	Helium core	2	2					
7	N	Nitrogen	14.54		2	3					
8	O	Oxygen	13.61		2	4					
9	F	Fluorine	17.42		2	5					
10	Ne	Neon	21.56		2	6					
11	Na	Sodium	5.14				1				
12	Mg	Magnesium	7.64				2				
13	Al	Aluminum	5.98				2	1			
14	Si	Silicon	8.15	Neon core			2	2			
15	P	Phosphorus	10.55				2	3			
16	S	Sulfur	10.36				2	4			
17	Cl	Chlorine	13.01				2	5			
18	A	Argon	15.76				2	6			
19	K	Potassium	4.34							1	
20	Ca	Calcium	6.11							2	
21	Sc	Scandium	6.56						1	2	
22	Ti	Titanium	6.83						2	2	
23	V	Vanadium	6.74						3	2	
24	Cr	Chromium	6.76						5	1	
25	Mn	Manganese	7.43						5	2	
26	Fe	Iron	7.90						6	2	
27	Co	Cobalt	7.86	Argon core					7	2	
28	Ni	Nickel	7.63						8	2	
29	Cu	Copper	7.72						10	1	
30	Zn	Zinc	9.39						10	2	
31	Ga	Gallium	6.00						10	2	1
32	Ge	Germanium	7.88						10	2	2
33	As	Arsenic	9.81						10	2	3
34	Se	Selenium	9.75						10	2	4
35	Br	Bromine	11.84						10	2	5
36	Kr	Krypton	14.00						10	2	6

Figure 3 Electronic structure of atoms (From Moore, C.E., *Atomic Energy Levels*, Vol. 2. Washington, D.C.: National Bureau of Standards Circular 467, 1952.)

Since electrons in the outermost band of an atom determine how an atom will react or join with a neighboring atom, the outermost band is called the valence band. Some electrons in the valence band may be so energetic that they jump into a still higher band and are so far removed from the nucleus that a small amount of impressed force would cause them to move away from the atom. Such electrons are responsible for the conduction of heat and electricity, and this remote band is called a conduction band. The difference in the energy of an electron in the valence band and the innermost subshell of the conduction band is called the band gap, or the forbidden gap.

Materials whose valence bands are full have very high band gaps (>3eV). Such materials are called insulators. Materials, on the other hand, that have relatively empty valence bands and may have some electrons in the conduction band are good conductors. Metals fall in this category. Materials with valence bands partly filled have intermediate band gaps (≤3eV). Such materials are called semiconductors (Figure 4). Pure semiconductors are called intrinsic semiconductors, while semiconductors doped with very small amounts of impurities are called extrinsic semiconductors. If the dopant material has more electrons in the valence band than the semiconductor, the doped material is called an *n*-type of semiconductor. Such a material seems to have excess electrons available for conduction even though the material is electronically neutral. For example, silicon has four electrons in the valence band. Atoms of pure silicon arrange themselves in such a way that, to form a stable structure, each atom shares two electrons with each neighboring atom with covalent bonds. If phosphorus, which has five valence electrons, is introduced as an impurity in silicon, the doped material seems to have excess electrons even though it is electrically neutral. Such a doped material is called *n*-type silicon. If, on the other hand, silicon is doped with boron, which has three valence electrons (one less than Si), there seems to be a positive hole (missing electrons) in the structure, even though the doped material is electrically neutral. Such material is called *p*-type silicon. Thus, *n*- and *p*-type semiconductors make it easier for the electrons and holes, respectively, to move in the semiconductors.

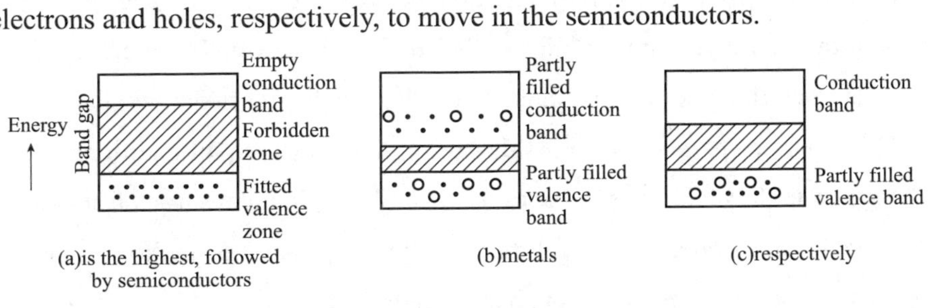

Figure 4 Electrical conduction is described in terms of allowed and forbidden energy bands. Band gap for insulators

3 *p–n* Junction

As explained earlier, an *n*-type material has some impurity atoms with more electrons than the rest of the semiconductor atoms. If those excess electrons are removed, the impurity atoms will fit more uniformly in the structure formed by the main semiconductor atoms; however, the atoms will be left with positive charges. On the other hand, a *p*-type material has some impurity atoms with fewer electrons than the rest of the semiconductor atoms. Therefore, these atoms seem to have holes that could accommodate excess electrons even though the atoms are electrically neutral (Figure 5). If additional electrons could be brought to fill the holes, the impurity atoms would fit more uniformly in the structure formed by the main semiconductor atoms; however, the atoms will be negatively charged.

p-n junction

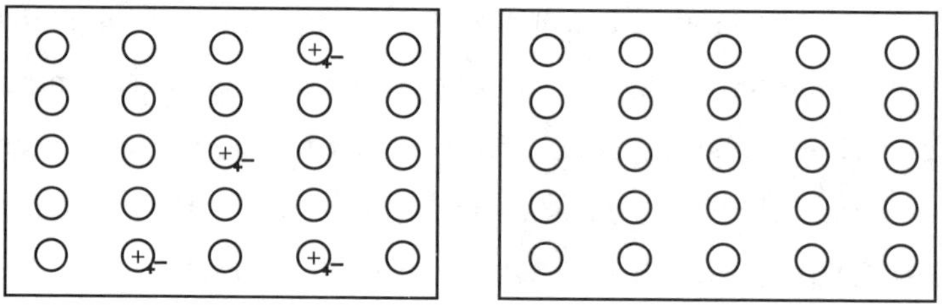

(a)*n*-type showing "excess" electrons as dots

(b)*p*-type showing "excess" positive holes as hollow circles

Figure 5 Representation of *n*– and *p*–type semiconductors

The above scenario occurs at the junction when a *p*-and an *n*-type of material are joined together as shown in Figure 6. As soon as the two materials are joined, "excess" electrons from the *n*-layer jump to fill the "holes" in the *p*-layer. Therefore, close to the junction, the material has positive charges on the *n*-side and negative charges on the *p*-side. The negative charges on the *p*-side restrict the movement of additional electrons from the *n*-side to the *p*-side, while the movement of additional electrons from the *p*-side to the *n*-side is made easier because of the positive charges at the junction on the *n*-side. This restriction makes the *p-n* junction behave like a diode.

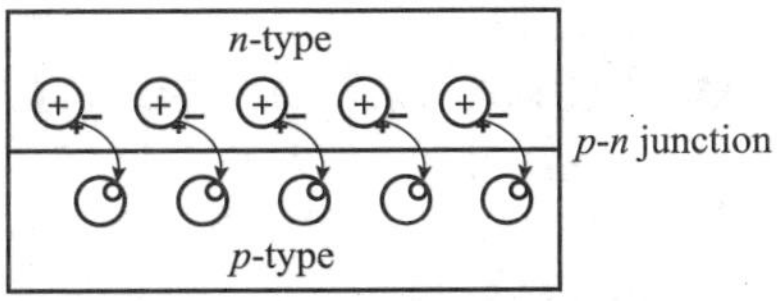

Figure 6 "Excess" electrons from n–material jump to fill "excess" holes on the *p*–side of a *p*–*n* junction, leaving the n side of the junction positively charged and the *p*–side negatively charged.

Selected from Chapter 9 of "*Principles of Solar Engineering* (third edition)", D. Yogi Goswami, CRC Press Taylor & Francis Group (2014).

Questions

(1) What is the principle of solar cell? Please explain what the photovoltaic effect is.

(2) What are the definitions of valence band and conduction band respectively?

(3) Please describe the definitions of conductor, semiconductor and insulator respectively.

(4) Please describe how the electron–hole pair generated.

(5) Is it possible that the conversion efficiency from radiative energy to chemical energy is 100%? Please state your opinion.

(6) There is energy loss of the semiconductor. Please describe the process that energy loss other than converting to electricity.

(7) What is *n*–and *p*–type semiconductor? Please give some examples of *n*–and *p*–type semiconductors respectively.

(8) Please describe how a *p*–*n* junction form and how it works.

Photocatalytic and Photoelectrocatalytic Water Splitting

1 Chemistry of Water Splitting

Water splitting reaction is an uphill reaction, which requires minimum Gibbs free energy of 237kJ/mol. The half-reactions are different depending on whether the electrolyte used is acidic or alkaline.

In alkaline electrolyte

$$4H_2O+4e^- \longrightarrow 2H_2+4OH^-$$
$$4OH^-+4h^+ \longrightarrow 2H_2O+O_2$$

In acidic electrolyte

$$4H^++4e^- \longrightarrow 2H_2$$
$$2H_2O+4h^+ \longrightarrow 4H^++O_2$$

2 Photocatalytic Water Splitting

Photocatalytic water-splitting

Natural photosynthesis in a plant was suggested as a cost-effective indispensable route for commercialization of solar water splitting. Researchers have devoted efforts to mimic these energy reactions since 1972 (Fujishima and Honda). When a semiconductor is suspended into water, **oxidation and reduction reactions of water** are performed at the interfacial region without any electrical work done, because the oxidation and reduction reactions are short circuited. In addition, both the conduction and valence band positions of the particulate material should be well matched with the hydrogen and oxygen evolution potential to split water (i.e., overall water splitting). Here, these microscale particles can catalyze water photolysis when their suspensions are exposed to the light. Besides the stability of the suspended particles, two additional qualities are required by the particulate, that is, catalysis (for water reduction and oxidation) and photovoltaic (light absorption, charge separation, charge transport) that enhance the complexity of the device [Figure 1(a)]. These functions are interrelated and jointly depend on the electrical, chemical, and optical properties of the semiconductors. The water-splitting reaction is significantly more endergonic, mechanistically complex, kinetically slower, and chemically more corrosive than the electron transfer-“only” processes. Although the photovoltaic electrolysis of water is highly efficient, the cost of hydrogen made by this route is quite high. Presently, efficiency and cost are the main challenges in the path scaling hydrogen production. In comparison with PV water cleavage, the particulate photocatalysis is a low-cost system. However, the separation of the explosive mixture of hydrogen and oxygen gas needs additional energy that reduces the overall **water–splitting** efficiency. Moreover, metal cocatalysts used for hydrogen and oxygen production from water are often good catalysts for the recombination of hydrogen and oxygen as well, which can reduce net water-splitting efficiency. Compared with thermochemical, PEC, and photobiological techniques of water splitting, photocatalytic water splitting is a promising technology to produce "clean" hydrogen. However, if this process is assisted by suspended photocatalysts directly in water instead of using photovoltaic and an electrolytic system, the reaction is in just one step; therefore, it can be more efficient. It has the following advantages: (1) reasonable **solar–to–hydrogen** (STH) efficiency; (2) low process cost; (3) the ability to achieve separate hydrogen and oxygen evolution during the reaction; and (4) small reactor systems suitable for household applications, thus providing for huge market potential.

Solar H_2 production-part I

Solar H_2 production-part II

The first recorded observations of a PEC phenomenon were made by Henri Becquerel (1839), who reported the photocurrent and photovoltage produced by sunlight acting on silver chloride-coated platinum electrodes in various electrolytes. Pure water absorbs only the infrared region of the solar radiation. In this region, photon energies are too low to simulate photochemical reactions. Therefore, water splitting is derived using solar and electrical energy support in the presence of semiconductors with small band gaps (1.6~2.5eV) with apt band positions or wide band gap semiconductors with a sensitizer or cocatalysts or both; the molecule or semiconductor is able to absorb sunlight and stimulate the photochemical reactions needed to drive the decomposition of water and lead to the generation of hydrogen and oxygen gases. During the photoelectrolysis of water, sunlight is absorbed by the isolated fabricated molecules (semiconductors) or biogenic materials (chloroplast or algae in a configuration coupled to a hydrogen-generating enzyme) or a hybrid of previous systems, in an aqueous electrolyte. These molecules are semiconductors or enzymes, either as a suspended particle in the water or as a macroscopic electrode unit in a PV cell or an electrochemical cell that is dipped in the aqueous electrolytes, used as an electron-hole relay. If we are in search of the apt photocatalytic material then we are really looking for characteristics such as high efficiency, good durability, or stability against **photocorrosion**, apt band gap, good lifetime of charge carriers, low concentration of defects, free path for charge carriers, low cost, and so on. It is really a challenge to find this in the single material. Besides an integrated device, high-tech systems are required for efficient photocatalytic water splitting or hydrogen production. Research and development activities on PEC materials, devices, and systems have made important strides in developing a strong synergy with contemporary research efforts in PVs, nanotechnologies, and computational materials. For commercial hydrogen production, the Nuclear Energy Research Initiative (NERI) set the PEC hydrogen production goals as follows: 15% conversion efficiency, 900hours replacement lifetime ($^1/_2$ year at 20% capacity factor), and ultimately synthesis via high-volume manufacturing techniques with a final hydrogen production cost of \$2.10/kg and \$300/m^2 PEC electrode cost.

Photoelectrocatalytic water splitting involves the conversion of light energy into a more useful energy product: electrical or chemical or both. Photoelectrocatalytic water-splitting device on absorption of solar light by one or more of the electrodes (photoanode and photocathode; at least one of the electrodes is a semiconductor) dipped in aqueous electrolyte that produce hydrogen and oxygen splitting of water. In principle, an internal electric field generated at the semiconductor-electrolyte interface uses the efficient separation of the photogenerated electron-hole pairs (excitons). Subsequently, photogenerated holes and electrons are responsible for oxidation and reduction of water, at the anode and cathode, respectively, for O_2 and H_2 production, as shown in Figure 1 (b).

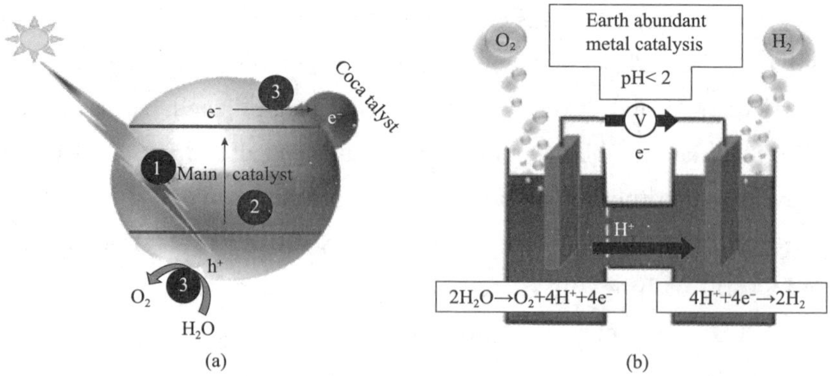

Figure 1 Illustration of the operating principle of the photocatalytic (①light absorption; ②charge separation; and ③charge transport).

Selected from Chapter 3 of "*Photochemical Water Splitting: Materials and Applications*", Neelu Chouhan, Ru-Shi Liu, Jiujun Zhang, CRC Press Taylor & Francis Group (2016).

3 Electrochemistry of Semiconductors at Electrolyte Interface

The anisotropic forces at the electrode-electrolyte interface and the charge transfer process occurring at the interface leads to rearrangement of electrons or holes inside the SC electrode and ions and solvent dipoles inside the electrolyte. The charge distributions on the electrode side are widely different for metals and semiconductors, in the following way:

Electrode-electrolyte interface

(1) Electrons and holes are the charge carriers in SC, whereas in metals, electrons are the only charge carriers.

(2) The carrier density is low in the case of semiconductors ($\sim 10^{14} \sim 10^{18}$/cc) than in metals ($\sim 10^{22} \sim 10^{24}$/cc).

(3) For metals, the charges are located at the surface, while for the semiconductors, the excess charge extends into the electrode throughout a significant distance (100~10000 Å). This region of potential drop is referred to as the space charge region.

The nature of the space charge layer depends upon the way in which charge transfer occurs across the interface. The band edges in the interior of the semiconductor also vary with the applied potential in the same way as the Fermi level. There are three different situations to be considered:

(1) At a certain potential, the Fermi energy lies at the same energy as the solution redox potential [Figures 2(a),(d)]. There is no net transfer of charge, and hence there is no band bending. This potential is referred to as the flat-band potential, E_{fb}.

(2) Depletion regions arise at potentials positive of the flat-band potential for an *n*-type semiconductor and at potentials negative of the flat-band potential for a *p*-type semiconductor [Figures 2(b),(e)].

(3) At potentials negative of the flat-band potential for an *n*-type semiconductor, there is excess of the majority charge carrier (electrons) in the space charge region, which is referred to as accumulation region [Figures 2(c)]. An accumulation region arises in a *p*-type semiconductor at potentials more positive than the flat-band potential [Figure 2(f)].

The charge transfer abilities of a semiconductor electrode depend on whether there is an accumulation layer or a depletion layer. If there is an accumulation layer, the behavior of a semiconductor electrode is similar to that of a metallic electrode, since there is an excess of the majority of charge carrier available for charge transfer.

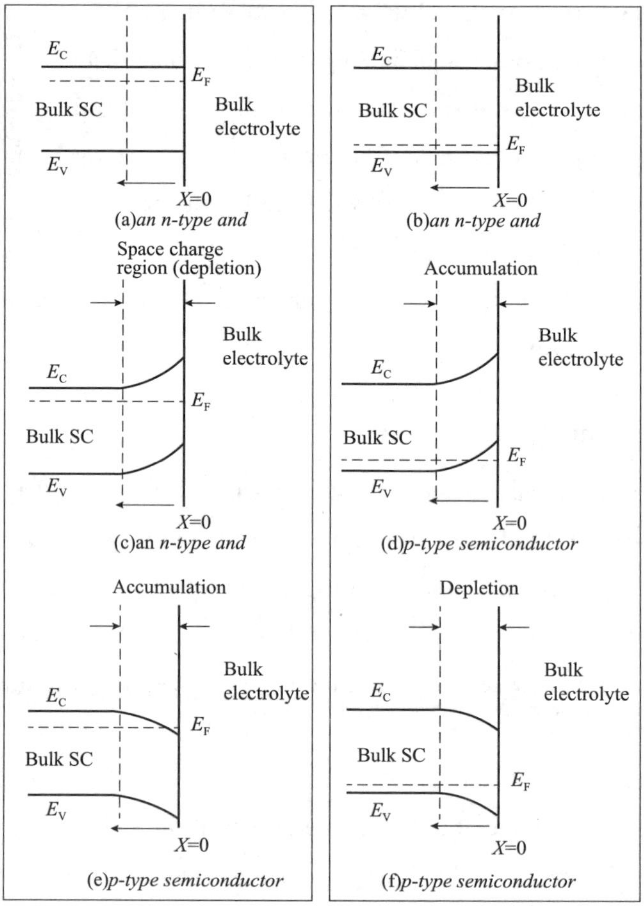

Figure 2 Effect of varying the applied potential (E) on the band edges in the interior of (a)~(c) an *n*–type and (d)~(f) *p*–type semiconductor

On the contrary, in case of depletion layer, there are few charge carriers available for charge transfer, and electron transfer reactions do not occur spontaneously. However, if the electrode is exposed to radiation of sufficient energy, electrons can now be promoted to the CB. If this process occurs in the interior of the semiconductor, recombination of the promoted electron and the resulting hole typically occurs, together with the production of heat. However, if it occurs in the space charge region, the electric field in this region will cause the separation of the charge. For example, for an *n*-type semiconductor at positive potentials, the band edges curve upwards and the hole moves toward the interface, and the electron moves to the interior of the semiconductor. The hole is a high-energy species that can extract an electron from a solution species, so that the *n*-type semiconductor electrode acts as a photoanode.

4 Electrolysis and Photoelectrolysis

In electrolysis of water, an external voltage of approximately 2.0V is required, which includes 1.23V thermodynamic energy requirement and polarization losses but in photo-assisted electrolysis of water, this requirement is reduced but still the external voltage is required which is called as bias voltage as shown in Figure 3. The semiconductor absorbs the light and produces electrons and holes. For an *n*-type semiconductor, electrons are moved to cathode, whereas holes oxidize water to produce oxygen. More than 100 materials and their derivatives have been explored till date to catalyze the overall water splitting or cause water oxidation or reduction in presence of an external redox mediator. However, none of them are capable of catalyzing overall water splitting having quantum efficiency greater than 10% has been reported.

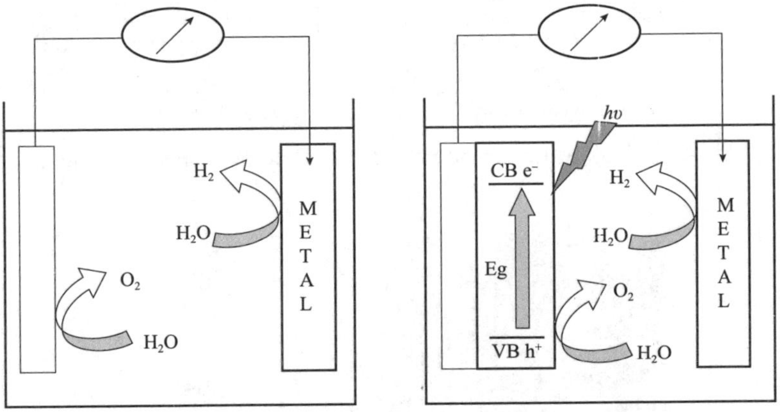

(a)electrolysis (b)photo-assisted electrolysis

Figure 3 Water splitting

Sacrificial reagents are often employed to evaluate the photocatalytic activity for water splitting (Figure 4). When the photocatalytic reaction is carried out in an aqueous solution including a reducing reagent, that is, electron donors or hole scavengers, such as alcohol and SO_3^{2-}, photogenerated holes irreversibly oxidize the reducing reagent instead of water. It enriches electrons in a photocatalyst, and a H_2 evolution reaction is enhanced. On the other hand, photogenerated electrons in the CB are consumed by oxidizing reagents, that is, electron scavengers such as Ag^+ and Fe^{3+} resulting in improved O_2 evolution reaction. These reactions using sacrificial reagents are studied to evaluate if a certain photocatalyst satisfies the thermodynamic and kinetic potentials for H_2 and O_2 evolution. These reactions are regarded as half reactions of water splitting and are often employed as test reactions of photocatalytic H_2 or O_2 evolution.

Selected from Chapter 3 of "*Visible-Light-Active Photocatalysis: Nanostructured Catalyst Design, Mechanisms and Applications*", Srabanti Ghosh, Wiley-VCH (2018).

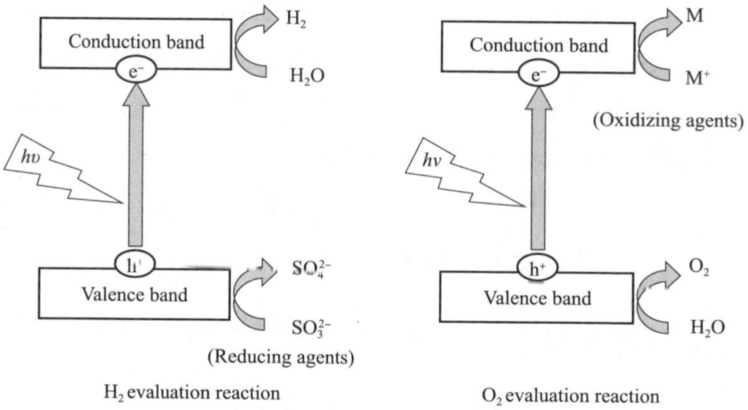

Figure 4 PEC water splitting in the presence of sacrificial reagents

Chapter 4

Biomass and Bioenergy (生物质和生物质能)

Find more information at HuaNeng, a public wechat account.

Part A Comprehensive Reading

Biomass Conversion to Chemicals

1 Background

What is Biomass?

Biomass has the potential to serve as a sustainable source of energy and organic carbon for our industrial society. The focus of this chapter is to provide a survey of different strategies to achieve chemical catalytic conversion of biomass-derived oxygenated feedstocks to value-added chemicals and fuels. The key reactions involved in the processing of biomass are hydrolysis, dehydration, isomerization, aldol condensation, reforming, hydrogenation/hydrogenolysis, and oxidation. Here, a few specific examples, namely efficient hydrolysis of **cellulose** over novel solid acids and synthesis of polyols by hydrogenation/hydrogenolysis of cellulose and sugar have been chosen for this review. Further, the selective conversion of platform molecules, such as furan, HMF, and biogenic carboxylic acids into intermediates, specialties, and fine chemicals has been considered. While many challenges are involved in biomass processing, understanding of fundamental reaction chemistry for different types of reactions can lead to the development of new approaches for specific processes.

Production of chemicals highly depends on non-renewable fossil resources such as petroleum, coal, and natural gas in modern society, which is widely acknowledged as unsustainable. The diminishing fossil fuel reservoirs together with ever-deteriorating environment associated with the worldwide fossil-fuel-based industry makes the use of alternative feedstock imperative. Biomass, which has utilized directly via combustion to produce heat for thousands of years, has been recalled for its renewable ability and enormous reserves. Biomass has been regarded as a sustainable alternative to fossil resources, and therefore, the effective utilization of biomass has greatly attracted the interest of the scientific and industrial communities.

The transformation of biomass to chemicals and fuels can be generally realized by three different techniques: thermal, biochemical, and chemical routes. Thermal techniques, like pyrolysis and gasification, can take full advantage of the entire organic substance of this resource. Nevertheless, these techniques still suffer from the disadvantages of low selectivity and high energy input which are unacceptable.

Bioconversion of biomass possesses the advantage of good selectivity, but sometimes suffers from low efficiency. Reasonable routes to promote biomass converted into high value-added chemicals under relative mild circumstance in liquid phase at a high selectivity are required.

2 Chemo–and Bio–strategy in Chemical Production: A Comparison

Fossil resources to chemicals

Chemical strategy has long been used in chemical industry, for the production of a wide variety of bulk chemicals from fossil (petroleum, coal and natural gas) and mineral resources. Chemical reactions and refining processes usually operate in various types of reaction vessels. In many cases the chemical conversions take place in special corrosion-resistant equipment at harsh conditions (elevated temperatures and pressures) with the use of catalysts. As different chemical reaction demands different catalyst and conditions, the whole process of chemical production usually requires multiple steps and reactions. A longer chemo-synthesis route often results in lower yield and higher cost of the produced chemicals. Moreover, unreacted substrates and by-products were often leaked as wastes to the environments, which have led to increased environmental protection regulations towards the chemical industry. For example, the leakage of hazards at the Union Carbide plant in Bhopal, India (1984) was the worst industrial disaster in the history of chemical production.

Comparatively, the developing bio-strategy has many differences with the mature chemo-strategy in chemical production (Table 1). (I) Bio-models often utilize the renewable biomass feedstock (mainly **carbohydrates** biomass) as substrates, which are more sustainable than utilization of the exhausting fossil resources. (II) Different from the chemo-strategy which has many non-natural synthesis routes, the **biosynthesis** or bioconversion routes for the objective chemicals usually exist in nature and only limited bio-routes have clear information for practical application; more bio-routes for important chemicals need to be explored to expand the category of bio-based chemicals. (III) The multistep natural bioconversion process (through specific enzymes) can be integrated in single cellular biocatalyst, which is incubated in one equipment. (IV) The bioconversion process usually takes place in specific equipment under bio-compatible conditions; the produced chemicals usually have low concentration in the bio-system and the low production efficiency leads higher capital investment (for unit production). (V) The bioprocesses are often cleaner than the chemical processes, using less energy and generating less hazard wastes.

Chemo- and Bio-strategy in Chemical Production

Table 1. Comparison of typical chemo- and bio-strategies in chemical production

	Chemo-strategy	Bio-strategy
Feedstock	Fossil and mineral resources	Renewable carbohydrates recourses
Production scale	Large scale	Medium and small scale
Category	Broader range	Limited range
Process	Multistep reaction, harsh condition, higher concentration	Integrated process, mild conditions, lower concentration
Technique	Well-developed techniques in synthetic chemistry and reaction engineering	Bio-synthesis techniques are in earlier stage of development
Environmental issue	More hazard waste and pollution	Less hazard waste and pollution

Collectively, both strategies have their own merits and bottleneck questions. The broad picture of bulk chemical production is: the chemo-strategy has significant advantages in economy and scale with mature, well-developed technologies, but is challenged by the exhausting resource and accumulating pollution; while bio-strategy is more sustainable and greener, but most of biochemicals still cannot compete with fossil-based chemicals due to the low production efficiency and the high capital investment.

3 Integration of Bio– and Chemical Processes

In line with the combined usage of a variety of nonrenewable and renewable resources, the conversion processes for the objective chemicals are become diversified as well. The bio-processes for some bulk chemicals and building blocks have been industrialized or near-term industrialized as they showed economical, technical and environmental beneficial towards the traditional chemical processes. However, bioprocesses themselves have many limitations. The most apparent one is the economic barrier in utilization of the **lignocellulosic**/cellulosic biomass. Separation of objective nutrients from the biomass is where major costs are generated. The difficulties in developing an efficient separating process are associated with the complexity of the structure. To improve the efficiency, the lignocellulosic/cellulosic biomass is usually pretreated and degraded by integration of chemical, physical and biological methods (Figure 1) if sugar substrates are needed for the downstream bioprocess to a variety of bio-based chemicals. Alternatively, biomass can be directly converted to biological oil, gas or specific materials via chemical process and modification.

Biomass to chemicals

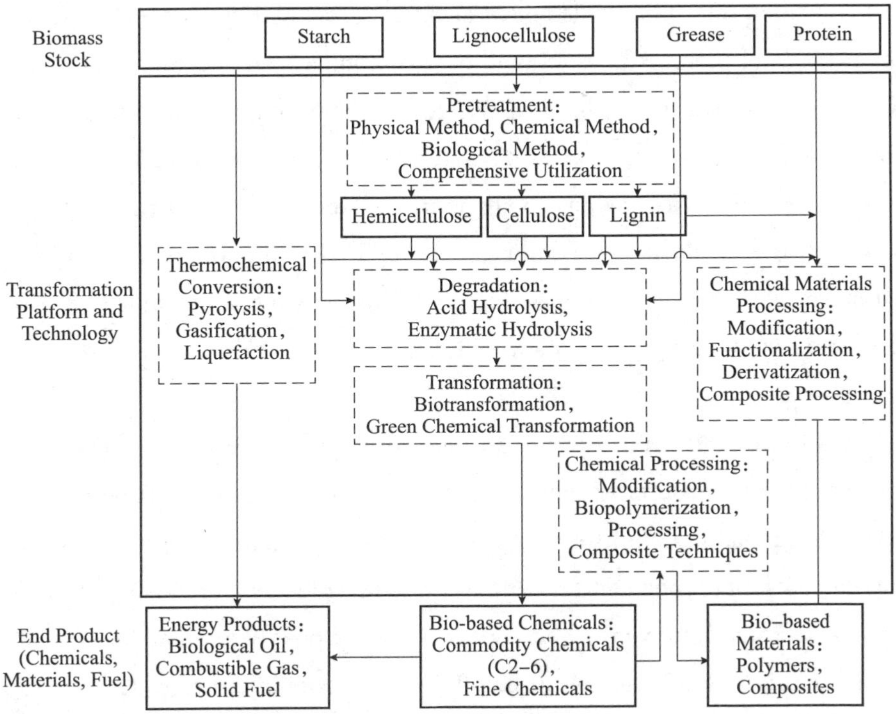

Figure 1 Integration of diversified bio- and chemo-processes in conversion of biomass into desired energy products, bio-based chemicals and materials

Another limitation of the bio-process is the range of the attainable chemicals: only limited chemicals have been built their feasible biosynthesis routes. For this reason, the bioprocesses are usually integrated with matured chemo-processes to expand the range of attainable semi-synthetic chemicals. These include poly lactic acids from bio-building block lactic acid; 1,4-butanediol, tetrahydrofuran, butyrolactone, maleic succimide, itaconic acid, and *N*-methylpyrrolidinone from bio-building block succinic acid; biodiesel, advanced alkane and alkene fuels from bio-derived fatty acids building blocks. Semi-synthetic processes can also be used in the production of value-added fine chemicals, such as the antimalarial drug artemisinin in which the precursor artemisinic acid is biosynthesized and then converted to the final product using organic synthesis.

4 Production of Polyols from Lignocellulosic Biomass

Biobased polyhydroxy compounds, including sugars, sugar alcohols, and glycerol can be converted into high value-added chemicals via catalytic conversion. However, unlike petroleum, biomass possess high oxygen content and specific carbon skeleton. In order to transform the biobased polyhydroxy compounds into high value-added chemicals, the selective removal of oxygen is the key step. Catalytic hydrogenolysis is an efficient way for biomass valorization. Here, we will provide a general overview of the catalytic conversion of polyhydroxy compounds into high value-added chemicals via hydrogenolysis. Figure 2 illustrates the value-added glycols with carbon chain in the range of 2~6, which can be derived from fatty acid esters and lignocellulose.

Cellulose, mainly from agricultural and forestry residues, is one of the world's largest organic raw material resource (nearly 40 billion tones production every year). Different from starch, cellulose cannot be digested by human. Therefore, instead of starch, human's main food source, cellulose was regarded as a promising resource that can be transformed into energy and chemicals in the post-petroleum era. To date, the one-pot chemical transformation of cellulose to polyols over heterogeneous catalyst is one of the most promising routes for the effective utilization of cellulose in a more efficient and convenient manner.

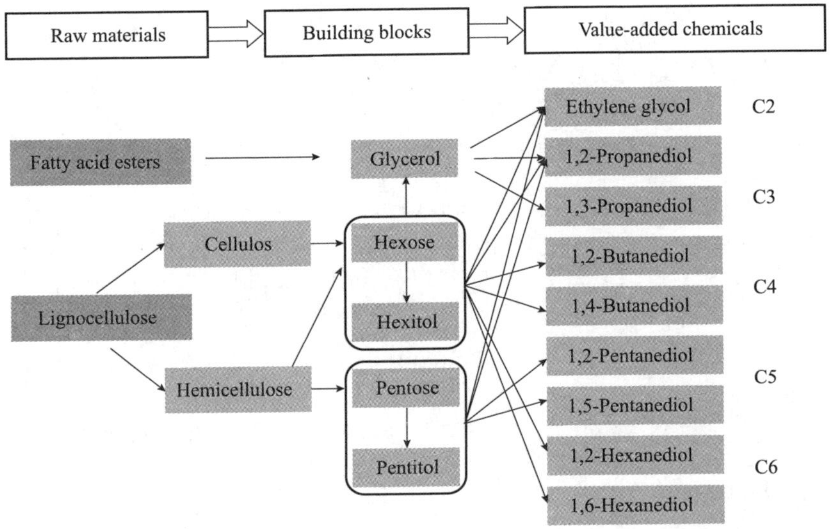

Figure 2 Value-added glycols with carbon chain in the range of 2~6 from fatty acid esters and lignocelluloses

The reaction network of cellulose hydrogenolysis is shown in Figure 3. Based on the progress made in cellulose hydrogenolysis, the real biomass was also used as starting materials for the production of polyols over similar catalysts used in cellulose conversion. However, researches still focus on the pure cellulose feedstocks because the real cellulose raw material contained some components that are difficult to transform or some elements that are poisonous to the catalysts. It is significant to spare additional efforts to investigate the influence of feedstocks in hydrogenolysis of cellulose.

The production of biodiesel

Glycerol is produced by hydrolysis of vegetable oil as a by-product of soap and by **transesterification** of oil and fats in the production of biodiesel. Due to the intensive interest in biofuels, the production of glycerol has increased enormously and the disposal of glycerol waste in biodiesel manufacture has limited the spread of the use of biodiesel. In this context, several ways of glycerol valorization have been described. One of the most promising ways to generate value-added products is hydrogenolysis especially 1,2-propanediol, 1,3-propanediol, and also EG (see Figure 4).

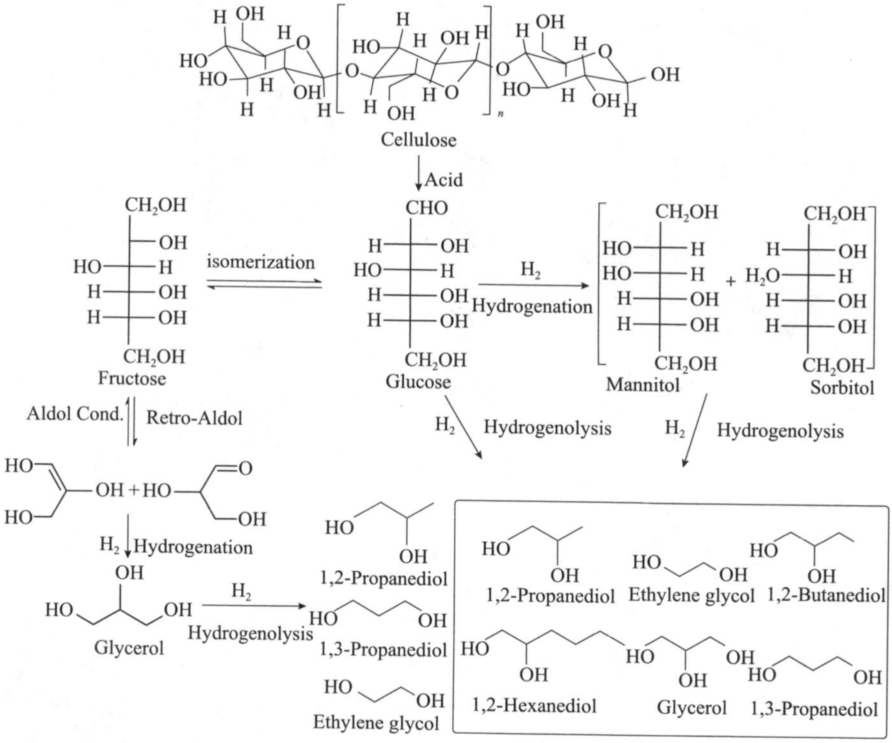

Figure 3 Typical pathways for the hydrolytic hydrogenation of cellulose to glycols

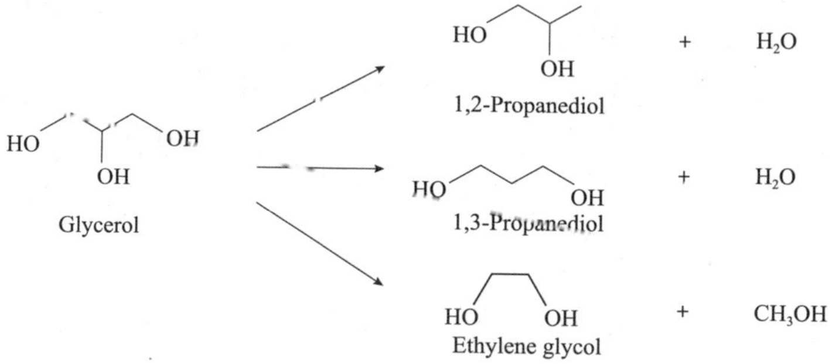

Figure 4 Conversion of glycerol to 1,2-propanediol, 1,3-propanediol and ethylene glycol

The hydrogenolysis of glycerol involves the dissociation of the C—C bonds and the C—O bonds followed by hydrogen addition. C—C hydrogenolysis of glycerol generates EG. On the other hand, in the C—O hydrogenolysis, 1,2-propanediol or 1,3-propanediol could be obtained depending on the catalyst used. The general challenge in the hydrogenolysis of glycerol lies in the selective cleavage

The conversion of glycerol

of C—O versus C—C bonds. Effective catalytic systems have been developed to selectively discriminate between C—C and C—O bonds. Cu-based catalysts were found to be selective for the formation of 1,2-propanediol with over 95% selectivity, while with the more active Ni and noble metal-based catalysts, both 1,2-propanediol and EG can be obtained due to the capability of cleaving both C—O and C—C bonds.

However, the selective hydrogenolysis of the secondary hydroxyl group in the cleavage of C—O bonds is a different problem. Several catalyst systems have been described which resulted in a significant selectivity for 1,3-propanediol. For example, WO_3 promoted Pt catalysts and ReO_x promoted Ir or Rh catalysts were especially effective in 1,3-propanediol formation among the reported materials.

Selected from Chapters 1 and 2 of "*Sustainable Production of Bulk Chemicals*", Mo Xian, Springer, 2015.

Notes

Cellulose: *noun.* (1) an insoluble substance which is the main constituent of plant cell walls and of vegetable fibers such as cotton. It is a polysaccharide consisting of chains of glucose monomers. (2) paint or lacquer consisting principally of cellulose acetate or nitrate in solution.

Carbohydrate: *noun.* any of a large group of organic compounds occurring in foods and living tissues and including sugars, starch, and cellulose. They contain hydrogen and oxygen in the same ratio as water (2 : 1) and typically can be broken down to release energy in the animal body.

Biosynthesis: *noun.* the production of complex molecules within living organisms or cells.

Lignocellulose: *noun.* a complex of lignin and cellulose present in the cell walls of woody plants.

Transesterification: *noun.* a reversible reaction in which one ester is converted into another (as by interchange of ester groups with an alcohol in the presence of a base).

Questions

(1) Currently, there are many disadvantages of fossil resources utilization. Please describe the reasons that people are seeking alternative renewable biobased resources.

(2) Why biomass attracted the interest of the scientific and industrial communities? What are the advantages of biomass resources utilization?

(3) Even though biomass resources have many advantages, there are still some problems hindered the industrialization of biomass conversion. Please describe what the disadvantages of biomass conversion are for industrial process?

(4) What techniques can be applied to transform biomass to chemicals and fuels?

(5) Biomass can be utilized by thermal techniques, like pyrolysis and gasification. However, there are disadvantages of this technique. What are they?

(6) Chemicals can be produced by both chemo- and bio-strategy. However, there are a lot of obstacles hindered the development of chemical strategy. What are the obstacles?

(7) Comparing with chemical strategy, what are the advantages of bio-strategy? What are the limitations of the bio-process?

(8) What are platform chemicals? Can you list all the platform chemicals?

Platform chemicals

(9) In the processing of biomass, what type of key reactions may be involved? Please give some examples of the industrial application of these reactions.

(10) What challenges may be involved in biomass processing?

(11) Cellulose, one of the world's largest organic raw materials resources, is regarded as a promising resource that can be transformed into energy and chemicals. Please describe the advantages of cellulose utilization. Besides, what are the bottlenecks of converting cellulose?

Part B Reading for Innovation

Cases for Bio–Based Raw Materials and Products

1 Converting Biomass to Ethanol

Plant cell walls are the most abundant form of biomass on the earth and thus an immense potential carbon source for bio-based products. Recent advances in biotechnology may now make it possible to exploit this raw material for the production of valuable **com–modities** such as ethanol(see Figure 1).

Biomass to ethanol

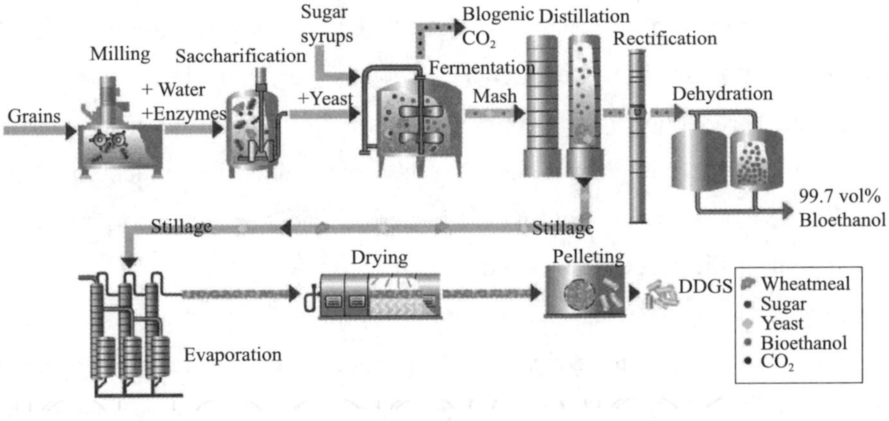

Figure 1 Schematic flow diagram of bio–ethanol production process

Plant cell walls are composed of crystalline **bundles** of cellulose embedded in a covalently linked matrix of hemicellulose and **lignin**. This complex polymeric structure poses a formidable challenge for solubilization and bioconversion. Dilute acid can hydrolyze (break down) hemicellulose at 140℃ to yield pentose (5-carbon) and hexose

(6-carbon) sugars. These simple sugars (predominantly xylose and arabinose with some glucose) are common substrates for bacterial **metabolism**. However, no naturally occurring organisms yet cultured can rapidly and efficiently convert both pentoses and hexoses into a single product of value.

Advances in genetic engineering have made it feasible to redirect the metabolism of simple sugars in certain bacteria so that they form no unwanted byproducts and efficiently channel key metabolites only into a desired end product. This approach was initially taken by Lonnie Ingram and colleagues at the University of Florida to create a strain of common bacterium, Escherichia coli, having an altered metabolism that diverts carbon flow to ethanol. The scientists inserted genes cloned from *Zymononas mobilis* into the chromosome of *E. coli.* These genes code for the enzymes pyruvate decarboxylase (which converts the intermediate pyruvate into ethanol) and alcohol dehydrogenase (which makes the conversion more efficient). Pyruvate decarboxylase binds pyruvate more tightly than the enzyme lactate dehydrogenase which, in unaltered E. coli, converts pyruvate to lactate.

Although seemingly straightforward, this experiment in metabolic engineering was based on a great deal of genetic, biological, and biochemical information resulting from years of effort by many researchers. Previous work had shown that the *E. coli* chosen for the “production” strain could metabolize all of the major sugar constituents of plant biomass. More important, tools for the genetic and biochemical manipulation of *E. coli* were available only because the bacterium had been subject to intense study, making it perhaps the best characterized of all bacteria.

Ingram and his colleagues have now gone beyond their initial work with *E. coli.* to engineer the cloned *Z. mobilis* genes into other bacteria such as Klebsiella oxytocia and Erwinia species. Unlike *E. coli*, these bacteria require less preliminary treatment of the cell walls because they contain additional enzymes that allow direct uptake of complex sugars (such as cellobiose and cellotriose) from plant cell walls. Erwinia strains also contain enzymes called endoglucanases that aid in the solubilization of lignocellulose.

The scientists’ success in engineering bacteria to produce a valuable commodity like ethanol is the first step. Further work is now needed to make these processes economically competitive with production of ethanol from petroleum-based materials.

2 Nature's Nylons

Common orb-weaving spiders spin as many as seven different types of silk fiber, each critical to the spider's survival. The fibers are made of proteins that are high-molecular-weight linear polymers exhibiting a broad range of mechanical and physical properties. Some silks are strong, tough framework filaments that support the web. Others are elastic filaments that absorb the kinetic energy of insects striking the web. Accessory filaments are produced that wrap captured prey or provide cocoon material. Silk fiber production imposes high energy and material demands on arachnids. Consequently, spiders have evolved to produce protein fibers that are highly efficient structures and one of the best high-performance natural materials.

Spiders use dragline filaments to control their movement in the wind and to form the primary load-bearing framework of the orb web. The combined strength and toughness of dragline filaments are unmatched by any existing man-made fiber: they have a tensile strength two to three times greater than steel and an elongation-to-break ratio approaching 30 percent. Dragline filaments offer an attractive benchmark for next-generation materials because of their exemplary physical and mechanical properties and also because they are processed at ambient temperatures from aqueous media.

The fibers in dragline silk are made up of glycine- and alanine-rich linear proteins containing oligopeptide units that function like the hard and soft segment units of conventional man-made polymers. The unique properties and processibility of dragline silk proteins result from the length and distribution of the protein segments as well as the amino acid sequence of each segment. Recombinant DNA technology has now made possible the development of bioengineered **analogs** that theoretically will perform as well or better than natural dragline filaments.

Several academic, government, and industrial laboratories in the United States are currently conducting research along these lines. The initial focus has been expression of protein polymers in microbial hosts, resulting in the successful production of several structural proteins in experimental quantities. Researchers also have prepared several fibers and films from de novo designs produced via fermentation. Early experience suggests that the economics of protein polymer production will need to improve before high-volume production and commercialization of these polymers become possible. Additional challenges will be determining how the constituent protein structures and biological spinning apparatus influence the transition from aqueous solution to insoluble fiber and finding ways to mimic the effects of spider anatomy and physiology at an industrial scale.

3 Genetic Engineering to Increase Starch Biosynthesis

Starch is the main storage carbohydrate in most plants. It is a major harvest component of several crops and thus directly affects yield. Industrial starch demand has increased dramatically over the past decade, primarily because of growth in the production of high-fructose corn syrups and bio-ethanol. In addition, various specialty starches, such as amylose and waxy **starch**, are being recognized for their superior material and nutritional properties as well as biodegradability(see Figure 2).

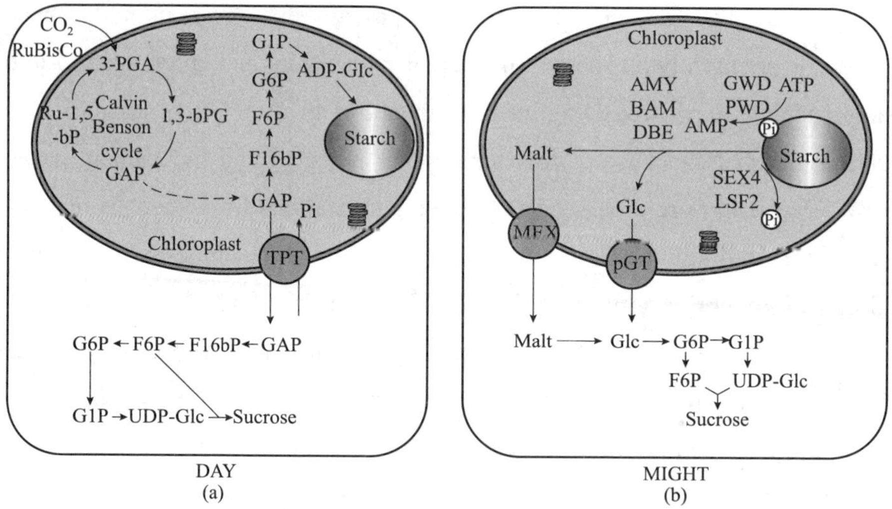

Figure 2 A schematic description of sugar synthesis at (a) day and (b) night times

Plant breeders have worked to increase the starch content of potatoes for a number of years; however, breeders typically do not achieve large changes. When large changes do occur, they prove difficult to work with since they can involve multiple genetic loci. The genetic engineering work described here involved a single gene and dramatically increased starch production, and thus harvested biomass per acre, without any apparent harmful effects on the plant.

The high-starch potato is one of the first genetic engineering products that targets industrial needs. Dry matter content (mostly starch) is the most important characteristic of potatoes for the processing industry because starch content affects processing cost, efficiency, and yield. For the food industry, high-starch potatoes are expected to improve efficiency and consume less oil during frying, thereby yielding fried products with more potato flavor, improved texture, reduced calories, and less greasy taste.

The strategy taken by Monsanto researchers was to increase starch content by enhancing the rate of starch biosynthesis. Starch biosynthesis in plants (and glycogen biosynthesis in bacteria) requires the enzymes ADPglucose pyrophosphorylase (ADPGPP), starch synthase, and branching enzyme. These enzymes pointbuild the large starch molecule. The ADPGPP catalyzes formation of ADPglucose from glucose 1-phosphate and ATP. The ADPglucose subsequently serves as the substrate for starch synthase. Plant cells tightly regulate the activity of ADPGPP, turning the enzyme "on" when excess carbohydrates are present and shutting it "off" when starch biosynthesis is not needed. The researchers reasoned that, since ADPGPP controls the amount of starch produced, addition of another enzyme that is not subject to control by the cell would cause more carbohydrate to flow into starch. Scientists at Michigan State University had utilized such an enzyme from the common enteric bacterium, Escherichia coli; this enzyme was previously discovered at the Pasteur Institute. The mutant enzyme GlgC16 causes *E. coli* to accumulate high levels of glycogen because the bacterial cells do not regulate its metabolic state (glycogen is similar to starch). The Monsanto group obtained the gene encoding GlgC16 designed it to be active only in tubers and transferred it into russet Burbank potato plants (the dominant potato variety in North America).

In some cases, tubers from transgenic plants containing the GlgC16 gene contained on average 25 percent to 30 percent more starch than tubers lacking the gene. Extensive U.S. field testing has shown the high-starch trait is stable in a number of different potato-growing environments. Furthermore, the trait has had no negative effect on plant growth or tuber yield, but, since the tubers contain more starch, the harvested dry matter per acre is substantially increased. The starch molecule in the transgenic potatoes, while more abundant, is structurally unchanged-important because the molecule's structure determines its end-use properties. The starch enhancement technology has now been extended to tomato and canola.

4 Biopolymers

The great majority of all biomass consists of natural polymers, and the great majority of all biomass is carbohydrate in nature. This means that the majority of all biomass is in the form of carbohydrate polymers, called polysaccharides. These natural polymers (biopolymers) can be used both as nature provides them and as the skeletal framework of other derived polymers. By far the most abundant of these carbohydrate polymers is cellulose, the principal component of cell walls of all higher plants. It is estimated that 75 billion tons of cellulose are biosynthesized and disappear each year, most of the disappearance being through natural decay.

Cellulosic plant materials are used as fuel, lumber, and textiles. Cellulose is currently used to make paper, cellophane, photographic film, membranes, explosives, textile fibers, water-soluble gums, and organic-solvent-soluble polymers used in lacquers and varnishes.

The principal cellulose derivative is cellulose acetate, which is used to make photographic film, acetate rayon, various thermoplastic products, and lacquers. The world's annual consumption of cellulose acetate is about 750000 tons, 400000 to 450000 tons being produced in North America. Cellulose acetate products are biodegradable.

While use of biopolymers, largely **polysaccharides**, as is and in modified form is now considerable, only an infinitesimal amount of that available is now utilized commercially in applications also served by petroleum-based polymers; so the potential is enormous. Broader application of such preformed polymeric materials awaits research and development.

5 Case Study of Lignocellulose Ethanol Processing

Feedstock supply and demand. Figure 3 shows a supply curve for corn stover. Supply curves identify the amount of a resource available in the market at a given price as determined by its value in the best alternative use. The supply price for a new crop typically consists of production costs (planting, care, harvesting) plus an allowance for land rent, where land rent represents the value of the land when it is used to produce another crop. The cost of corn stover may be lower than new crops, at least for some levels of use; stover and corn grain are produced together, making recovery of land costs for corn stover unnecessary because they have already been accounted for in grain profit calculations.

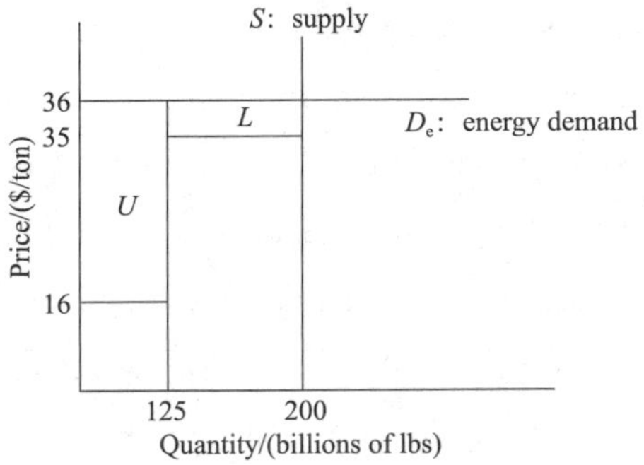

Figure 3 Corn stover supply and demand curve

The present economic value of corn stover arises from two sources. First, erosion and fertilizer requirements are reduced when corn stover is left on the ground. Second, corn stover replaces low-grade hay when fed to cows. A first approximation of the stover supply curve facing ethanol processors is a step function (Figure 3). It begins horizontal at net harvest cost (*U*). Processing plants in well-chosen locations in cash-grain areas could acquire stover that is not used by livestock producers at slightly above harvest cost. The second step of the curve is the higher value that livestock producers are willing to pay for stover used as feed (*L*). If processors are willing to pay slightly more than the livestock value, all stover supplies would be diverted to industrial uses. The supply curve is vertical where industry uses all available supplies, provided that the amount of land that is planted with corn is kept fixed.

If the new cellulose conversion technology develops successfully, energy products will create a new market for corn **stover**. The horizontal demand curve, D_e in Figure 3, shows energy producers' returns for processing a unit of corn stover (the difference between energy product price and stover processing costs). The market determines corn stover use up to where processors' returns equal corn stover harvest costs-the intersection of supply and demand. If energy prices rise, the demand curve will shift upward, and progressively more corn stover (and other raw materials) will be available at progressively higher prices in a market economy. Estimates of corn stover use opportunity costs to determine the dollar per ton (height) steps of the corn supply curve.

Table 1 gives net harvest cost estimates based on harvest expenses and fertilizer replacement costs for the midwestern United States. Harvest costs are estimated from hay harvesting costs (fixed machinery replacement costs and variable operating costs). The calculations use a harvested stover tonnage that includes an amount (30 percent) left on the field for soil conservation compliance. The fertilizer estimate is based on replacing phosphorus and potassium contained in the harvested corn stover. The stover net harvest cost is $16.50 per ton. Reported production costs for several energy feedstocks in a Midwest location are considerably higher than harvest cost for corn stover. The estimates range from $42.00 per ton for sweet sorghum to $95.00 per ton for canary grass in Ames, Iowa, where crop rents are relatively high. Costs were lower at Chariton, Iowa, due to reduced land costs, ranging from $36.00 for sweet sorghum to $93.00 per ton for alfalfa. The feed value of hay can be similarly calculated using some adjustments for total digestible nutrients and the protein deficiency of corn stover in comparison to hay. The value of corn stover as a feed is about $35.00 per ton based on the 1994 hay price.

Table 1. Costs of Corn Stover Harvest in the United States, 1993

	Direct harvest costs				
Operation	Reported fixed cost / $	Fixed cost / ($/ton)	Reported variable cost / $	Variable cost / ($/ton)	Total / ($/ton)
Rake	2.43/acre	1.22	1.52/acre	0.76	—
Baler	3.14/bale	6.28	2.05/bale	4.10	—
Total	—	7.50	—	4.86	12.36

	Indirect harvest costs		
Fertilizer	Application rate /(lbs/acre)	Price /($/ton)	Total / ($/ton)
P_2O_5	13	150	0.49
K_2O	71	206	3.66
Total	—	—	4.14

The volume of corn stover available to the processing industry can be approximated using estimates of available corn stover, cattle population and forage requirements, and the availability of hay for forage. Calculations (not shown) involved multiplying a state's corn area by a stover yield estimate that leaves an allowance for compliance with the Conservation Reserve Program. Similarly, the estimate of cattle feed demand is the product of cattle population and forage requirement per animal, less hay supply for each state. The industry supply is the supply less feed demand—in this case about 125 billion pounds (62.5 million tons) available at low prices near the harvest cost of $16.00 per ton. At prices above the feed cost of $35.00 per ton, the entire stover supply of about 200 billion pounds (100 million tons) would be available to the processing industry.

Approximately 7.5 billion gallons of ethanol could be produced at about $0.46 per gallon from corn stover. An additional 4.5 billion gallons of ethanol may be produced at higher costs due to a potentially higher price for corn stover above the feed cost of $35.00 per ton. When correction is made for the relative fuel efficiency of ethanol and gasoline, the 12 billion gallons of ethanol from corn stover translates to the equivalent of 9.6 billion gallons of gasoline or about 9 percent of annual U.S. gasoline consumption (about 110 billion gallons).

Transportation costs. A large processing plant could exploit economies of scale but would also require vast amounts of corn stover (about 2.9 million tons for a 350 million gallons per year ethanol plant). Given the average corn density in Iowa, for example, all the corn stover available in a 50 miles radius would be required. Since the delivered cost of the stover increases with distance, the transportation component of input cost could become large and offset economies of scale when such a bulky material is used as a feedstock. The firm's average input costs (AICs) can be approximated by the formula AICs=P_0 + 2tr/3, where P_0 is the harvest cost, r is the radial distance from the plant (in miles), and t is the transportation cost (in dollars per ton per mile). Based on quotations from Iowa trucking firms, the 1994 transportation rate is between $0.10 and $0.15 per ton-mile (for shorter distances). Hence, the average total cost of all stover drawn from within 50 miles is $20.00 to $21.50 per ton.

The lower stover estimate was used to calculate processing cost (below) because the routine of a large plant could reduce the short-haul rate and because the average corn stover density may understate availability at the local level. Some offsetting bias may exist in these calculations; distance is measured “as the crow flies” instead of on a particular road network and therefore may be understated. Also, the average corn density may understate availability at the substate level because it combines low-density crop production from cattle grazing areas of the south with high-density corn production in cash-grain areas of the north. Well-chosen plant locations would be in the cash-grain areas, where stover could be acquired at near harvest costs. Overall, the calculations do suggest that transportation costs are not a major barrier to operation of a large-scale plant, at least in the Corn Belt.

Processing costs. Revised material flows and cost estimates are shown in Table 2. This analysis adjusts reported cost data on bioprocessing for uniformity and for comparisons that indicate tradeoffs by decision-makers. The cost data also conform to the procedures of Donaldson and Culberson (1983) that facilitate comparison to petrochemical processes. Variable costs include materials and utilities. Labor costs are included where available. Fixed costs are limited to capital costs, calculated as the annual payment on a 15-year, 10 percent interest, fixed annual payment mortgage on the entire plant cost. Other **expenditures** are excluded from fixed costs, including overhead (because there is no opportunity cost) and insurance (in order to offset risks of the profit stream, which should be considered elsewhere). Expenses such as labor for plant maintenance or taxes on the plant could be included, but individual situations vary, and data were not uniformly available. Cost data for some petrochemical processes (styrene, ethylene, ethanol) were developed using Donaldson and Culberson's estimates of input requirements, yields, and plant costs—combining input requirements with recent price data to estimate material and utility expenditures, updating capital expenditure data with a price index for plant and equipment, and giving annual payment for a 15-year mortgage. These calculations assume that stover is available at near its harvest cost of \$16 per ton and includes a transportation cost. It is also assumed that technology for fermentation of pentose sugars is fully implemented.

Other cost adjustments were required for conversion to corn stover from wood chips. In particular, the lignin content of corn stover is only one-half of that for wood chips, so the one-half of electrical plant capacity that was sold as a byproduct credit is removed. Also, the 10 percent gasoline mixing operation was eliminated, so estimates now refer to a pure ethanol basis. Input prices and capital outlays were indexed to a 1993 basis. Finally, the feedstock cost is corn stover harvest cost adjusted for transportation as discussed above. The overall production cost for ethanol is estimated at $0.46 per gallon (refer to Table 2).

Table 2. Production Cost Estimate for Plant Processing Corn Stover to Ethanol

	Units per 10^3 gallons of ethanol	1993 price /($)	Units expenses per 10^3 gallons of ethanol
Wood/stover, ton	8.3	19.805/ton	164.38
Sulfuric acid, pounds	297.0	86.20/ton	12.80
Lime, pounds	219.0	40/ton	4.38
Ammonia, pounds	470.4	200/ton	47.04
Nutrients, pounds	18.1	230/ton	2.08
Corn liquor, pounds	63.3	266/ton	8.42
Corn oil, pounds	3.9	413.4/ton	0.81
Glucose, pounds	37.0	258/ton	4.77
Catalyst	1000	0.01/unit	10.00
Disposal, ton	0.34	20/ton	6.80
Water, million gallons	19.87	0.002/gallon	0.04
Total Materials	247.92		
Labor, man–year	41.0	198,800/year	3.49
Foremen, man–year	9.0	34,000/year	0.88
Supervisors, man–year	1.0	40,000/year	0.11
Total Materials + Labor	252.40		
Capital Allowance	211.4		
Total Production Cost	463.8		
Total Processing Cost (net of stover)	299.4		
Ethanol output	350 million gallons		
Input requirement	2.90 million tons		
Plant cost	$562 million		

Fuel efficiency. A study in 1996 reported that ethanol reduces **fuel efficiency** in automobiles by 2 percent when used in the standard 10 percent blending proportions with gasoline. Hence, a gallon of ethanol is worth only 80 percent of a gallon of gasoline to an automobile engine. Thus, for $0.46 per gallon of ethanol, it will cost $0.46 (0.80=$0.58) to produce the ethanol equivalent to a gallon of gasoline.

6 Softening Wood the Natural Way

A new experimental process called "biopulping" uses natural lignin-degrading fungi to "soften" wood prior to pulping. Wood usually is pulped either chemically, by modifying and dissolving the lignin, or mechanically, by grinding to separate the fibers. Studies of biopulping have generally examined it as an "add on" to mechanical processes, although it shows promise in combination with chemical processes as well. Biopulping decreases the energy required for mechanical pulping by 30 percent to 50 percent, results in stronger paper, and is environmentally sound. Electrical energy is the major cost associated with mechanical pulping, and an energy savings of even 25 percent by addition of the fungal process would justify substantial investment in the latter method.

Research on biopulping has resulted from an eight-year collaboration among the U.S. Department of Agriculture (USDA) Forest Service's Forest Products Laboratory, the University of Wisconsin's Biotechnology Center, the University of Minnesota, and industry. Only a few of the several hundred fungal species and strains examined thus far show promise; the relatively uncommon *Ceriporiopsis subvermispora* appears to be superior. This fungus species is effective with aspen and loblolly pine, two woods used extensively in mechanical pulp production. Biopulping requires about 2 weeks and causes little damage to cellulose. Brief steaming of the wood chips enhances the process because it gives the fungus a competitive advantage over indigenous microorganisms, and supplementing the fungal mycelium with corn steep liquor reduces the quantity (cost) of the initial fungal inoculum. Modest rates of aeration can control the self-heating associated with biopulping.

As presently envisioned, commercial biopulping would involve steaming, cooling, and inoculating the wood chips and then allowing incubation with aeration for about two weeks prior to pulping. A disadvantage of biopulping is that it darkens the chips, requiring increased chemical use during pulp bleaching. An advantage is that biopulping largely destroys pitch components (lipophelic constituents of wood, e.g., oils, resins, fatty acids) that cause problems during the pulping process.

A pitch reduction process similar to biopulping has already been commercialized. The fungus in this case, *Ophiostoma pilferum*, does not degrade lignin. When inoculated into wood chips being piled for storage, the fungus selectively degrades the pitch. A white strain of *O. pilferum* used in a product from the Sandoz Chemicals Corporation also acts as a biocontrol agent, keeping out dark-colored fungi that normally discolor wood chips during storage.

Selected from "*Biobased Industrial Products: Priorities for Research and Commercialization*" Committee on Biobased Industrial Products, Board on Biology, Commission on Life Sciences, National Research Council. National Academy Press, 2000.

Important Words and Phrases

Commodities: *noun.* a raw material or primary agricultural product that can be bought and sold, such as copper or coffee.

Bundle: *noun.* a collection of things or quantity of material tied or wrapped up together. *verb.* (1) tie or roll up (a number of things) together as though into a parcel. (2) push or carry forcibly.

Lignin: *noun.* a complex organic polymer deposited in the cell walls of many plants, making them rigid and woody.

Metabolism: the chemical processes that occur within a living organism in order to maintain life.

Analog: *adjective.* (1) of, relating to, or being a mechanism or device in which information is represented by continuously variable physical quantities of or relating to an analog computer. (2) of, relating to, or being an analogue.

Starch: *noun.* (1) an odorless, tasteless white substance occurring widely in plant tissue and obtained chiefly from cereals and potatoes. It is a polysaccharide which functions as a carbohydrate store and is an important constituent of the human diet. (2) powder or spray made from starch and used before ironing to stiffen fabric or clothing. (3) stiffness of manner or character. *verb.* stiffen (fabric or clothing) with starch.

Stover: *noun.* the dried stalks and leaves of a field crop (especially corn) used as animal fodder after the grain has been harvested.

Expenditure: *noun.* the action of spending funds.

Fuel efficiency: *noun.* the capacity of an engine, especially that of a vehicle, to obtain energy from fuel.

Part C Reading for Fun

New Sustainable Plastics

1 Motivation

To develop sustainable plastics, researchers are creating polymers that not only have excellent properties during their useful life, but also offer options for end-of-life management. A life cycle assessment approach should underpin these developments, which should improve the efficiency of polymer production and reduce energy inputs. In some contexts, it may be beneficial to use waste plastics, biologically-derived feedstocks or even CO_2 as raw materials to make new plastics. Ideally, renewable feedstocks should be suitable as "drop in" substitutes in existing chemical processes.

2 Introduction

The vast majority of plastics on the market today are made from petrochemicals, which are ultimately derived from oil and gas. Continuing to use these raw materials will require a reduction in both energy use and carbon emissions associated with these processes. These reductions need more research into improved manufacturing and processing methods. In the longer term, it seems likely the world will shift from a manufacturing system based predominantly on fossil fuels, towards one that makes products from a range of sources including recycled chemicals and renewable raw materials, using processes powered by renewable energy. Chemistry can also help to develop routes to make plastics that recycle waste CO_2 as a co-monomer and lock it into the polymer backbone.

Most common petrochemical-derived polymers such as polyethylene and polypropylene take many years to break down in the environment. These molecules are carbon chains that typically do not contain chemical groups that could act as obvious 'break points' for chemical or biological degradation, and in many cases are water resistant. The polymer chains can assemble to form crystalline regions that confer excellent thermal and mechanical properties, making the material extremely durable in use, but also less amenable to degradation after use.

Only a handful of polymers have been designed to be easily degradable in the environment—for example the polyester polybutylene adipate terephthalate (PBAT), which is used as a compostable packaging material because it has similar properties to low-density polyethylene (LDPE) but reacts with water to fully degrade. There are almost no commercial polymers that have been explicitly designed to make recycling easier. Plastics need to be designed for recycling and the area of depolymerization chemistry needs fundamental exploration, theoretical understanding and exploitation. One option is to exploit the benefits conferred by biologically sourced monomers, which tend to be more oxygen rich than hydrocarbons, and to use these oxygen-bearing chemical groups to provide "break points" for degradation. Another option is to exploit non-permanent interactions between chains to produce polymers that can disassemble, or crystallites that can be disrupted with a lower energy cost.The Figure 1 shows the global plastic production reached 359 million tons in 2018.

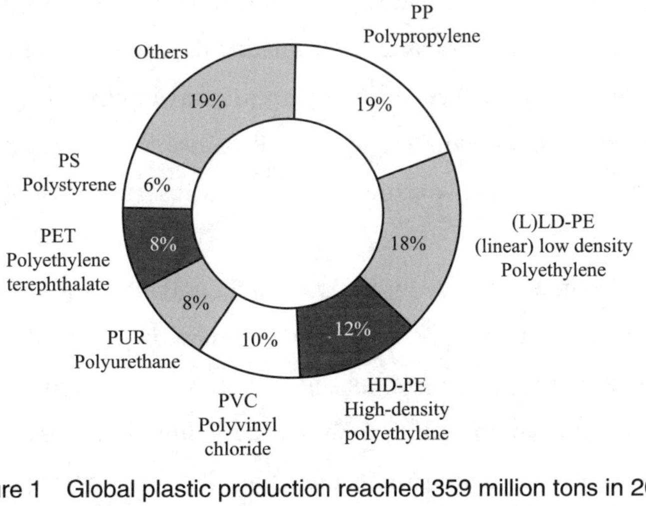

Figure 1 Global plastic production reached 359 million tons in 2018

Designing new plastics is important for sectors as varied as packaging and construction, and specialized plastics are also needed in many other sectors, including electronics, transportation, furnishing, leisure and agriculture (see Figure 2). These areas may accelerate uptake of new sustainable plastics and help new materials secure a foot-hold in the market.

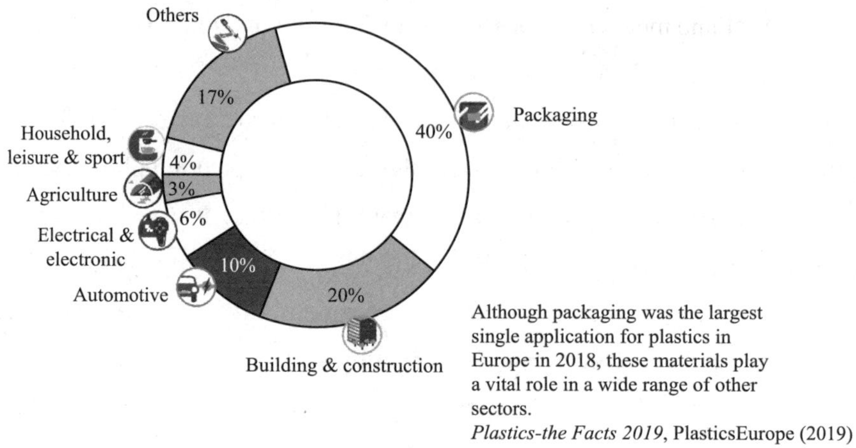

Figure 2 The uses of plastics

3 Polymers from Renewable Resources

Bio-based polymers currently make up a tiny proportion of the plastics market. Polylactic acid (PLA, also known as polylactide) is one of the most widely used bio-based synthetic polymers, found in products such as disposable cutlery and plastic cups. Thanks to the aliphatic ester chemical groups in the polymer chains, PLA can be completely broken down by industrial composting processes (see Figure 3). Life cycle assessments suggest that PLA has lower overall CO_2 emissions compared with conventional petrochemical plastics such as polyethylene and polypropylene.

PLA is most commonly made by a ring-opening polymerization of lactide, a cyclic diester made from lactic acid. This starting material is produced by the fermentation of plant sugar extracted from crops such as sugar cane. Work is ongoing to utilize waste lignocellulosic feedstocks, including food and agricultural wastes, in order to make lactic acid production even more sustainable. Other lactones and cyclic monomers can be polymerized either to make variants of PLA or other aliphatic polyesters, and these approaches allow for some control over the polymers' chemical, physical, thermal and mechanical properties.

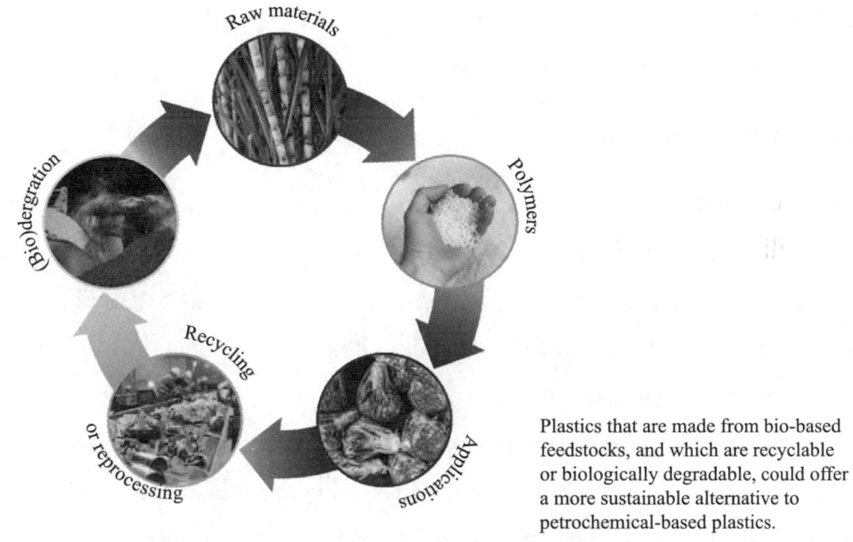

Figure 3 A sustainable cycle for plastics

Other bio-based polymers can be produced using monomers such as terpenes, including pinene derived from pine tree oil, limonene oxide derived from waste citrus fruit peel, and succinic anhydride.

Researchers have developed highly active and selective catalysts for these (co-) polymerization reactions, but further improvements are needed to translate these to sustainable industrial processes that produce plastics with more desirable characteristics such as higher mechanical strength and better barrier properties.

It is important to note that simply because a monomer can be produced from renewable resources does not guarantee improved sustainability. For example, bio-based polyethylene terephthalate (PET) can be produced by making the ethylene glycol co-monomer from glucose, but this gives rise to impacts such as eutrophication and acidification. The other co-monomer for PET, the aromatic diacid, cannot currently be economically produced from renewables. Alternative aromatic monomers derived from sugars show promise to produce fully bio-based polymers with similar properties to PET, but there are still challenges with monomer and polymer production. Better catalysts and processes could help, for example by using high pressure CO_2 together with new catalysts to reduce polymerization temperatures and energy consumption.

It must be stressed that the material properties of a polymer are decisive for its applications. The thermal and mechanical properties of PLA, for example, define its possible applications, and exclude many others.

4 Using Waste Molecules

Better reuse of waste molecules, either from industrial, agricultural or other chemical processes, will be important in reducing the carbon emissions associated with plastics. But CO_2 can also be used, in conjunction with a co-monomer, as a building block for plastics. In many ways, it is an ideal feedstock—it is cheap, abundant, and is the waste product from many industrial processes.

For example, polycarbonates can be produced by reacting epoxides with CO_2. This process is operating at commercial scale, and life cycle assessments have quantified the reductions to overall greenhouse gas emissions. The CO_2 copolymerization process is critically dependent upon the catalyst, and recent inventions have uncovered the opportunity to increase performances using mixed-metal catalysts, for example of magnesium and zinc. As chemists develop a better understanding of how these metals cooperate, that should help to develop better catalysts that can make the process even more efficient, reducing manufacturing costs and prompting wider adoption of these plastics.

It is also possible to add carbon monoxide to ethene during its polymerization, which introduces carbonyl groups into the polyethylene chain. These can act as convenient break points, so that chemical reactions involving enzymes or light could degrade the polymer molecules in the environment after a few months. Such materials are already being used in some plastic packaging, but it takes extremely high pressures to drive the polymerization reaction that produces them, raising manufacturing costs and making the process very energy intensive. So chemists are developing catalysts and methods that can make such polymers under much more moderate conditions, which could offer a more convenient route to such degradable plastics.

In the longer term, chemistry has a key role to play in generating completely new and disruptive technologies for plastics, potentially replacing some carbon-based monomers with new materials based on other elements. For example, sulfur is a large scale waste from the petrochemical industry and finding ways to recycle it into polymers could be useful. Polythioesters exploit sulfur to introduce convenient break points in the polymer structure that could be targeted by future recycling processes.

Natural polymers such as cellulose, chitin and starch can serve as renewable and environmentally degradable resources, either to make monomers or as materials in their own right. Cellulose is a major component of natural plant fibers, and the world's most abundant natural polymer. It is already used to make a range of products including cellophane film, textiles and packaging.

But processing cellulose can be challenging. Cellulose molecules are held together by networks of hydrogen bonds, which means that cellulose does not melt and does not easily dissolve in common solvents. Chemists are developing a range of strategies to make cellulose processing easier, for example by exploiting ionic liquids to dissolve cheaper low-grade cellulose feedstocks such as corn husks or wheat straw. This can reduce the cost or number of steps involved in cellulose processing. Finding new ways to fine-tune the physical properties of cellulose derivatives may help to open up applications such as low-cost packaging, or to avoid the use of additives such as plasticizers(see Figure 4).

Figure 4 Rolls of cellulose-based plastic film

Significant research effort is devoted to the efficient transformation of lignocellulosic wastes into monomers for plastic production. This research effort benefits from expertise in heterogeneous catalysis, engineering, bio-processing and synthetic biology. It also allows for the production of polymer-natural fiber composites and for the improvement of natural materials like paper by producing polymer composites.

5 Conclusion

In future a range of recycled and reused chemical raw materials should be used to make polymers. These raw materials can and should include the products of chemical recycling of existing plastics. Because polymers are manufactured at large scale, the decisions about which monomers to deploy are guided to a large extent by feedstock availability and price. Techno-economic and sustainability assessment should be integrated at an appropriate stage of technology translation and ensure that fundamental research is well targeted.

The infrastructure used to make and process polymers represents a large capital investment that should not be abandoned in the short term. Thus, the development of sustainable “drop in” substitutes for the highest-volume plastics (like polyethylene and polypropylene), which could use the same infrastructure, is a research priority.

When designing polymers, sustainability assessment over the entire materials life cycle is important. For example, future materials should be designed to minimize toxicity and environment impacts. As chemists find new ways to turn renewable molecules and recycled wastes into polymers it is important to design those polymers so that they can ultimately break down to environmentally non-persistent biochemicals and products. Moreover, any new plastics will need to be easier and less energy intensive to recycle. Research is needed to improve understanding of how polymer structure and microstructure affect their physical properties, recycling and degradation chemistries.

Selected from “*Science to enable sustainable plastics*”A white paper from the 8th Chemical Sciences and Society Summit. London, UK, 2020

Logistics of Lignocellulosic Feedstocks: Preprocessing as a Preferable Option

1 Component variations

With biorefineries based on renewable plant resources, the biomass composition, fiber structure, and molecular weight of the components differ from species to species. For instance, the lignin content decreases in the order of softwoods, hardwoods, and grasses. Furthermore, variations can also be found within the same biomass category depending on the species and genotype, the plant part, the physiology of the plant, its location, cultivation conditions such as different fertilizer treatments, and the harvesting time. This diversity influences both the suitability of plants as a feedstock for bioenergy conversion, as well as the process and supply chain design. For instance, the structural analysis of nine crops (Miscanthus, switchgrass, fiber sorghum, fiber corn, spelt, tall fescue, cocksfoot, hemp, and Jerusalem artichoke) showed that Miscanthus has the highest content of cellulose, hemicellulose, and lignin, while the fiber corn shows the highest starch content. Moreover, the later the harvesting date, the lower the content of proteins and mineral compounds in the crops. This effect is caused by a decrease of the non-carbohydrate-rich part of the plant leaf with age, while the related amount of the carbohydrate-rich stem increases. Consequently, late winter harvesting provides a biomass that has a higher content of structural components (cellulose, hemicelluloses, and lignin) compared to an early harvested crop. In fact, the higher total carbohydrate content makes this feedstock more suitable for biorefineries. Nevertheless, harvesting in autumn is recommended more strongly due to the higher total dry matter yield, which outweighs the effect of the carbohydrate content.

2 On-Site Measurements of Biomass

To implement optimal pretreatment and value-adding processes, easy and rapid analytical methods are needed to characterize the physical and chemical composition of diverse feedstock—preferably on-site. The chemical composition of the native and pretreated biomass is nowadays usually analyzed according to established laboratory analytical procedures, so-called National Renewable Energy Laboratory (NREL) Analytical Procedures (LAP), and American Society for Testing and Material (ASTM) procedures. Many pretreated feedstocks have already been characterized using these procedures and are available on the Biomass Feedstock Composition and Property Database. The most frequently used method is based on the total acid hydrolysis of the biomass sample and allows the quantification of the hemicellulose, cellulose, and lignin amount. However, these reactions typically require harsh conditions and analysis takes place only on a laboratory-based scale. To supply biorefineries with suitable feedstock, easier analytical methods need to be established for feedstock characterization.

Infrared spectrometry is widely used to characterize agricultural products. Unlike conventional techniques, near- and mid-infrared spectrometry allows the fast analysis of biomass composition and provides quantitative information within a short time period without any previous sample preparation or degradation. These techniques can be used in off-line and on-line modes and are suitable tools for feedstock characterization. In general, spectrometry has often been applied to characterize the morphological and chemical composition of biomass.

If near-infrared spectrometry is applied to analyze a biomass composition, the calibration should be qualitatively or quantitatively modelled by multivariate data analysis. The chemometric analysis can be done based on principal component analysis (PCA) and various regression algorithms using specific software. The PCA is usually applied to preliminarily decrease the high data density. Through its application, orthogonal directions of maximal variance and the relationship between variables and objects can be identified. The focus of the PCA is on the determination of qualitative information and identification of the relationship between the absorption and concentration of components in multicomponent samples. To correlate the concentration of individual components with the measured absorption and to quantify the single components, multiple linear regression (MLR) and multiple regression (MR) or a partial least squares regression (PLS) algorithm can be used afterwards. The advantage of the PLS algorithm compared to the other regression methods is that it can utilize all of the information of the whole spectral data set. It accounts for all correlations and can describe them using only a few components. Using this method, multiple variables can be calculated simultaneously. Additionally, the wavelength regions where analytes of interest absorb can be identified by the use of specific algorithms. This method simplifies the analysis and improves the calibration quality.

3 Harvest and Pre–Transport Treatments

Despite their abundant supply, readily available biomass streams are rarely allocated homogenously and uniformly. A lignocellulosic biomass from different feedstock origins, such as agricultural and forest residues and dedicated energy crops, often cannot be transported and further processed in an efficient and cost-effective way due to the physical characteristics and complexity of cellulosic biomass. Thus, the supply chain for raw biomass distribution requires several preliminary steps in order to maintain low overall costs. In Figure 1 the delivery system of feedstock to biorefineries as the intended end-users is depicted. These include the collection and harvest of biomass and the subsequent storage or pre-processing of the raw material. Treating the biomass for transport is necessary in order to improve the flowability and stability of the raw material.

Depending on the infrastructural conditions and the distribution policy, different strategies for the handling and transportation of biomass can be applied. General pre-processing methods are readily carried out on-field or after transportation into storage facilities, conversion plants, or in small satellite processing facilities. The main approach for the transportation of the provided feedstock takes place by using existing distribution channels, such as highways or railroads. Depending on the local conditions and the type of transported material, waterways or multi-link transport chains are also used as routes for transportation. Spatial proximity of the location of harvest to the processing facility is preferable since it reduces both the cost of handling (e.g., loading and unloading) and transportation. As the transportation costs of biomass contribute the major costs in logistics, a reasonable strategy for the supply must be worked out. In novel approaches, two logistic scenarios have been distinguished: the conventional bale system (CBS) and the advanced uniform supply system (AUS).

The CBS can be designated as a local depot of field-dried and baled biomass, which supplies biorefineries, without changing the properties or stability of the biomass. In contrast, AUS pre-processing depots lead to a final uniform material that can be easily stored and transported into biorefineries. Although the latter involves pre-processing, it still remains more cost-effective for larger biorefineries (up to 10000 dry metric tons per day) in comparison with the conventional system, and higher biomass volumes can be employed. In CBS, the capacity per hour of loading and unloading of biomass from vehicles into depots is limited(see Figure 1).

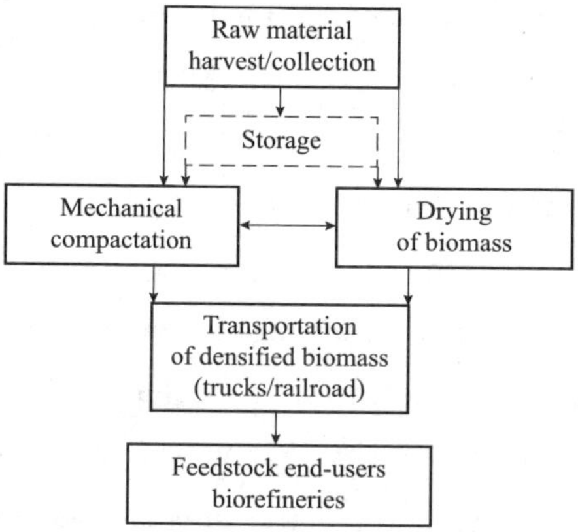

Figure 1 Graphic representation of the common supply chain from raw biomass to biorefineries

Regardless of the selected facilities and transportation modes for the supply and delivery of feedstock, a pre-transport treatment remains essential for most lignocellulosic biomass, with the possible exception of grass and herbaceous plants that have a very low lignin content. Residual lignocellulosic materials are usually found to have low bulk and energy densities and high moisture content. The limiting factor of transportation and shipping is therefore often restricted by the large volume of a given biomass as opposed to weight alone. Depending on the type of feedstock used, the volume may alter. Accordingly, an excerpt of reported bulk densities found in the literature is given in Table 1.

Table 1. Raw materials with corresponding bulk densities

Raw material	Bulk density/(kg/m^3)
Grass and crop residues (loose)	70

(continued)

Raw material	Bulk density/(kg/m^3)
Hardwood chips	230, 402
Miscanthus	350
Softwood chips	180~190
Straw (loose)	20, 36, 24~111
Switchgrass (loose)	49~3231

Mechanical compactation. In order to make raw biomass transportable and reduce the overall costs, feedstock should be densified before loading. Mechanical compactation of bulky, uneven, or fluffy biomass especially simplifies handling, increases bulk density, and leads to a product with uniform properties. Biomass packing can be distinguished by applied forces by means of briquetting, extrusion, palletization, or tabletizing, resulting in increased bulk densities and characteristic shapes of the densified biomass (e.g., cylindrical, cuboid, pillow-shaped, round). Biomass is mechanically compressed between a roller press, by a screw/piston, or with the aid of a perforated and rotating hard steel die. These methods are usually additionally combined with techniques for size reduction such as cutting, grinding, or milling. Common on-field processing includes balling of grasses and crop residues by balers, or using mulchers for woody biomass. In Table 2 bulk densities of raw biomass after compression are given.

Table 2. Bulk densities of selected raw biomass after densification

Treatment	Straw, bulk density/(kg/m^3)	Switchgrass, bulk density/(kg/m^3)
Baled	110~200	142~186
Briquettes	—	480~530
Chopped	20~80	—
Cubed	320~670	—
Hammer milled	20~110	115~182
Pelleted	560~710	—
Tapped	68~323	34~130

Drying. The drying of biomass is a common practice in on-field handling and harvest of biomass, and makes storage and transportation more convenient.

In agriculture, this procedure has been successfully approved in order to prevent degeneration and microbial attacks of biomass. The operation involves different procedural steps, which include mowing of biomass, windrowing, balling, collection, staging, and finally storage of biomass. Typically, stored dry bales must have a moisture content of less than 20% in order to prevent biological degeneration. Depending on the chosen bale shapes (rectangular or round) and the duration of storage, different amounts of dry matter losses might occur, which will have a significant influence on feedstock quantity and quality. Moreover, field-drying is associated with some problems, including harvest timeline, seasonal changes, and risk of rehydration, which influence the overall costs and the degree of dryness. Over-intensive drying might result in irreversible shrinkage of pores and reduction of accessible surface area for enzymatic degradation. Mild drying at room temperature only leads to a decline of small pores, which are already inaccessible for enzyme deconstruction and do not hamper sugar conversion. The concept of drying is often directly combined with mechanical compactation of biomass in the form of pelleting or briquetting. Thus, there are different recommendations for optimal moisture content for the densification procedure: for wood residues an optimal moisture content of 8% was determined, where high-density and long-term performance was seen during compactation. In the case of straw, usually a dry moisture content of 15% or less is used, while recently researchers found that briquettes produced at a low moisture content of 9% yielded the maximum densities of $700kg/m^3$ for wheat, oat, canola, and barley straw. In a parametric study, researchers found that the pelletizing temperature (125~180℃) and size, as well as the moisture content (0~10%) had an influence on the pellet strength and dimensions of compacted spruce. Using a pressure of 300MPa for 5s, the biomass density could be increased to $1000\sim1260kg/m^3$. Additionally, dry biomass is also appropriate for the production of pellets in combination with thermal pretreatment such as torrefaction. It has been shown that dry torrefaction of woody biomass leads to a more thorough removal of hemicellulose and to more char combustion reactivity than wet torrefaction.

4 Decentralized Value–Adding Options

Bag–hydrolysis. One of the major obstacles in using lignocellulosic biomass as a source for product generation is the recalcitrance of biomass followed by the necessity of a pretreatment step. Currently, most lignocellulosic raw materials are therefore mechanically or thermochemically pretreated, followed by enzymatic hydrolysis. The available carbohydrate content of raw materials is between 56% and 74% and can be converted into mono- and disaccharides by enzymatic hydrolysis. Typically, enzymatic hydrolysis is conducted in stirred tank reactors, which might not be feasible for use in decentral processing units, since higher investments costs are necessarily paired with high operating costs due to stirrer power consumption. Enzymatic hydrolysis at high solid loadings (>15%) increases the process economics, but leads to higher viscosities and yield stress coupled with poor rheology. The prerequisite for a cost-effective biomass conversion is a reactor design that allows a maximal conversion of cellulose with a minimal amount of enzymes at high solid loadings.

On a laboratory scale, different modes of saccharification such as shaking, gravitational tumbling, and hand stirring have been investigated and have shown that effective initial mixing facilitates high conversion rates, due to sufficient enzyme distribution. As part of this, new reactor concepts are currently developed that focus mainly on the employment of new or combined stirrer models or the use of existing solid-state bioreactors for high solid loadings with maximal working volumes of 3~4 kg. For an on-site hydrolysis biomass operation, single-use bag reaction systems have been considered. These are relatively simple reaction vessels made of polyethylene (0.2mm thick) that can be mixed by simple rocking movements. As a substrate, beech wood or crude beech wood cellulose fractions are used up to solid concentrations of 10% (w/w) with low enzyme dosages. In Figure 2, the temporal change of liquefication of pretreated biomass during enzymatic hydrolysis in a single-use bag system is depicted. After only 8h of hydrolysis, degradation of previous fiber structures is already apparent. With increasing hydrolysis time, full liquefication and visible loss of solid matter is reached. The rocking motion of the bags in combination with enzymatic conversion leads to homogenization of the slurry.

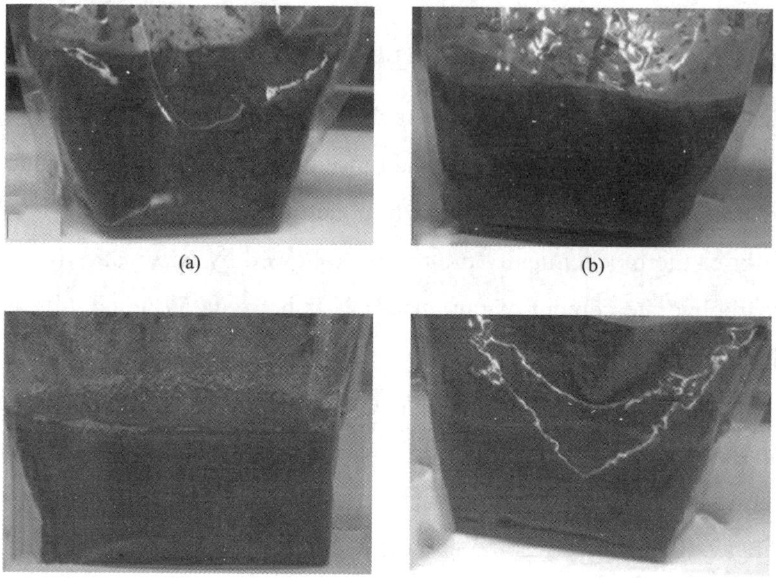

(a) (b) (c) (d)

Figure 2 Enzymatic hydrolysis in single-use bag systems, with a crude cellulose solid loading of 100g/L. Duration of hydrolysis: (a)3h, (b)8h, (c)20h, (d)30h. Temperature: 50℃

Figure 3 shows the course of hydrolysis during enzymatic conversion of the pretreated biomass. The utilization of a single-use bag system allows the enzymatic conversion of pretreated beech wood into glucose and xylose, reaching maximum concentrations of 48g/L glucose from cellulose and 13g/L xylose from the remaining hemicellulose, respectively. Direct application on-farm seems to be possible, since a low-cost polyethylene foil can be easily prepared and welded for use with simple and minor equipment costs. Moreover, further truck transportation while enzymatic conversion takes place is conceivable if the trucks are equipped with systems comparable to concrete mixers. Mixing by stirring has been shown to be an adequate way of mixing in enzymatic hydrolyses. Hence, it needs to be further examined whether this approach is economically competitive. This concept leads to further conversion potentials such as simultaneous transportation and saccharification, which has already been discussed for direct pipeline application. First findings indicated that transport of biomass slurries from wood residues and wheat straw in pipelines is possible at approximately 30% biomass loadings with regard to rheology.

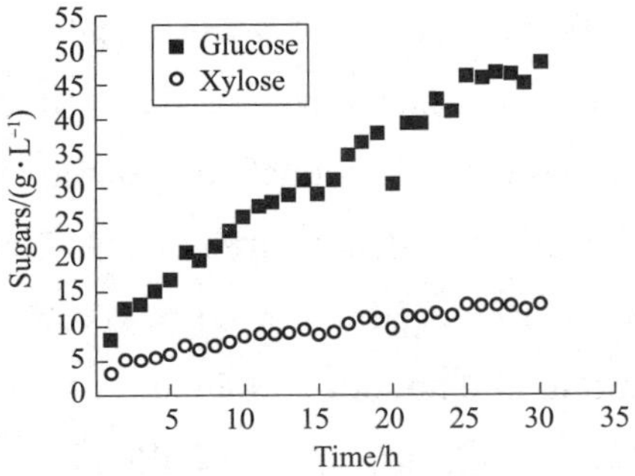

Figure 3 Glucose and xylose concentrations during enzymatic hydrolysis at a crude cellulose (with Hemicellulose Impurities) solid loading of 100g/L using single-use bags. Enzymes: 6% cellulase, 6% xylanase. Rocking-motion: 30rpm. Temperature: 50℃

Ensiling of Biomass and Product Generation. A common method for long-term preservation of medium moisture content lignocellulosic plant material, such as from grassland, can be provided by ensiling the biomass. The resulting silage is the product of solid-state lactic acid fermentation, which can be used as a high-quality animal feed the entire year. During storage of fresh plant material under anaerobic conditions, the production of organic acids, as well as a pH shift, takes place. This prevents the growth of other microorganisms and the decomposition of the plant biomass. Other additional benefits compared to dry storage are that biomass loss during handling is reduced and there is no need for costly pre-drying of the biomass. Additionally, ensiling can be considered as a combination of storage and pretreatment of biomass. Direct on-farm realization of both storage and pretreatment makes handling considerably easier, since pretreatment is conducted at ambient temperature and pressure without the need for chemical or thermal pretreatment, leading to cost and energy savings.

Nowadays, ensiling is also suggested as a biological pretreatment for lignocellulosic biomass and has been reported to improve the enzymatic saccharification of biomass. In several studies using pretreated silage biomass, anaerobic fermentation using the yeast *Saccharomyces cerevisiae* or *Kluyveromyces marxianus* was conducted separately from the ensiling process allowing for conversion of available sugars into ethanol. On a laboratory scale, a combined approach of ensiling and simultaneous saccharification and fermentation with the addition of cellulase and glucoamylase for 20days resulted in 6.5wt% ethanol (a yield of 169g/kg dry mass) using 250g non-sterile forage paddy rice plants. This indicates that a reaction of lactic acid fermentation and subsequent ethanol fermentation coupled with the enzymatic deconstruction of plant material is possible. Thus, a decentralized ethanol production system that can be directly applied in the on-farm level was proven on a laboratory scale. In a further study, this system was extended and directly applied on-farm. For this purpose, rice plants were chopped, harvested, and baled, resulting in 0.8m tall round bales of 1 m diameter with a weight of 273~283kg (48%~50% dry weight). Before baling, a mixture of commercially available enzymes and microorganisms for lactic acid fermentation and ethanol production was dissolved in 40kg of distilled water and added to the feedstock. In detail, cellulase (0.74~0.77FPU/g DM) of *A. celluloslyticus* and a *glucoamylase* from *Aspergillus niger* (0.29g/g DM) was employed as well as freezedried lactic acid bacteria (2×10^5cfu/g DM) and freeze-dried *S. cerevisiae* (3×10^6cfu/g DM). The bales were then wrapped with a plastic film for ensiling. The resulting effluent was collected by further entrapment of the bale system with a water-impermeable polyethylene plastic foil. The entire solid-state fermentation was monitored directly on-field without temperature control by sample collection from different locations of the bales. The effluent collected at the bottom of the bales was recovered monthly and ethanol was recovered with a vacuum distiller. During the operating period of 1~6 months, 90.9~139.6g/kg DM of ethanol was produced during Simultaneous Saccharification and Fermentation (SSF) in whole round bales, which corresponds to a maximum yield of 14wt % ethanol.

Selected from "*Biorefineries*" Kurt Wagemann, Nils Tippkötter Editors. Springer, 2019. Nils Tippkötter, Sophie Möhring, Jasmine Roth, and Helene Wulfhorst.

Appendix Ⅲ. The 17 Sustainable Development Goals
（联合国17个可持续发展目标）

Appendix Ⅳ. Principles of Green Chemistry & Engineering
（绿色化学与绿色工程原则）

The 12 Principles of Green Chemistry

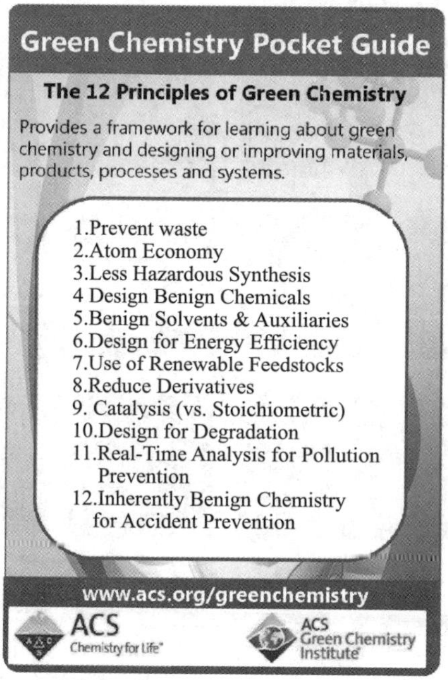

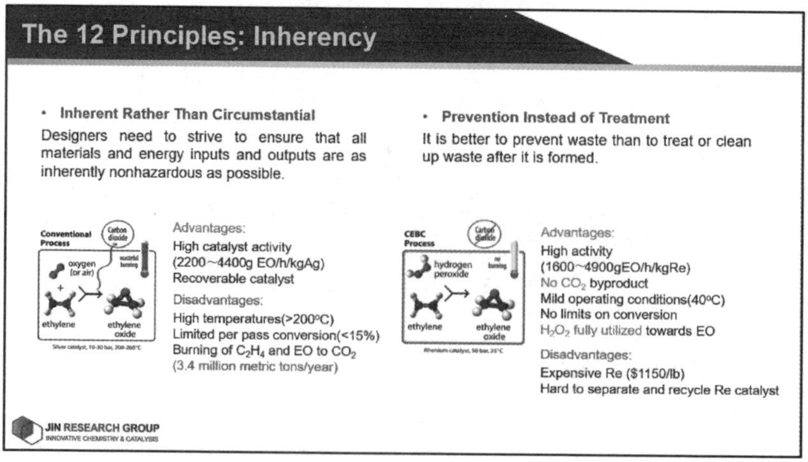

The 12 Principles: Simplicity

- **Conserve Complexity**

Embedded entropy and complexity must be viewed as an investment when making design choices on recycle, reuse, or beneficial disposition.

- **Meet Need, Minimize Excess**

Design for unnecessary capacity or capability (e.g., "one size fits all") solutions should be considered a design flaw.

- **Minimize Material Diversity**

Material diversity in multicomponent products should be minimized to promote disassembly and value retention.

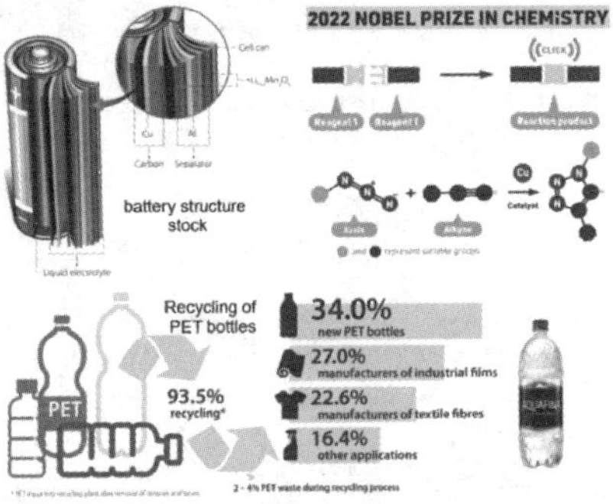

JIN RESEARCH GROUP
INNOVATIVE CHEMISTRY & CATALYSIS

The 12 Principles: Output

- **Output-Pulled *vs* Input-Pushed**

Products, processes, and systems should be "output pulled" rather than "input pushed" through the use of energy and materials.

- **Design for Commercial "Afterlife"**

Products, processes, and systems should be designed for performance in a commercial "afterlife."

The diaper structure

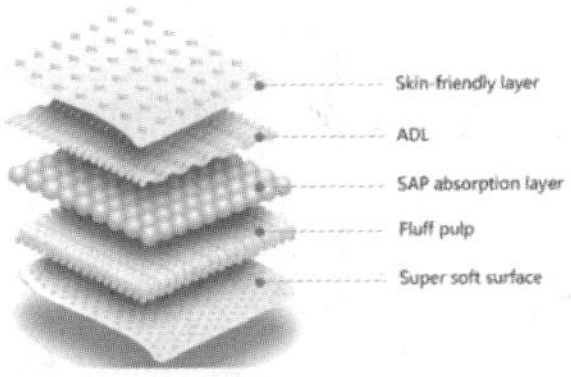

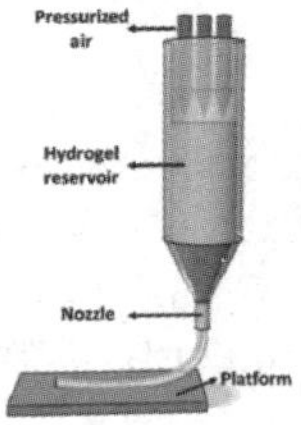

3D Printing

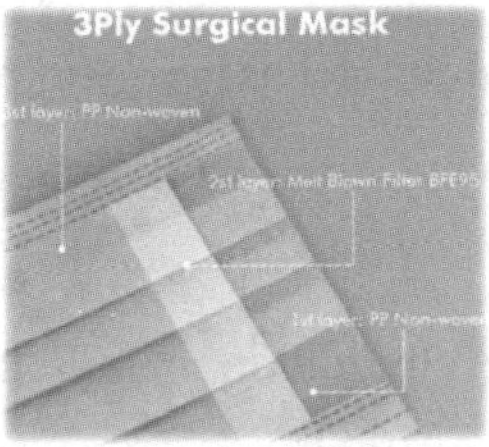

The structure of a mask

JIN RESEARCH GROUP
INNOVATIVE CHEMISTRY & CATALYSIS

The 12 Principles: Efficiency

- **Maximize Efficiency**

Products, processes and systems should be designed to maximize mass, energy, space, and time efficiency.

- **Design for Separation**

Embedded entropy and complexity must be viewed as an investment when making design choices on recycle, reuse, or beneficial disposition.

- **Integrate Material and Energy Flows**

Design of products, processes, and systems must include integration and interconnectivity with available energy and materials flows.

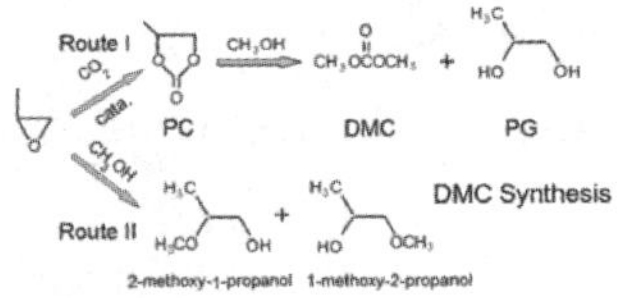

DMC Synthesis

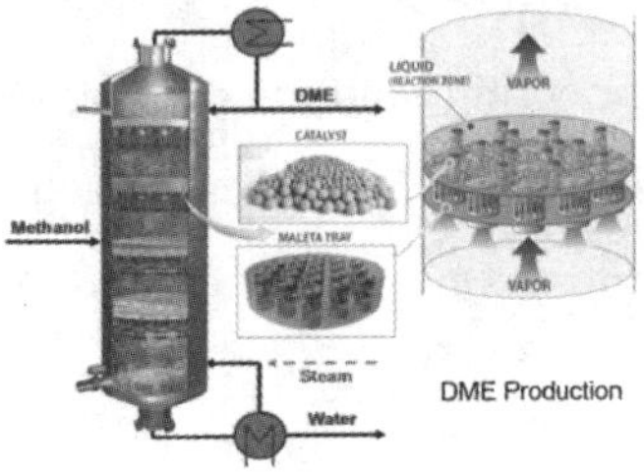

DME Production

JIN RESEARCH GROUP
INNOVATIVE CHEMISTRY & CATALYSIS

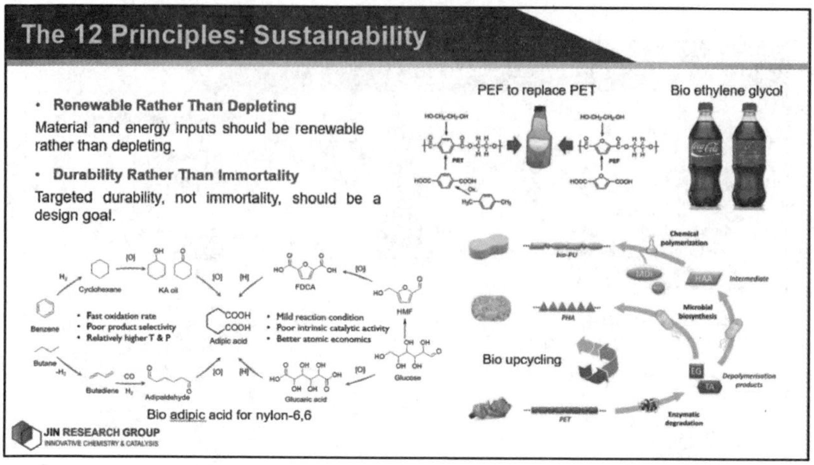

Nine Principles of Green Engineering

(1) Engineer processes and products holistically, use systems analysis, and integrate environmental impact assessment tools.

(2) Conserve and improve natural ecosystems while protecting human health and well-being.

(3) Use life-cycle thinking in all engineering activities.

(4) Ensure that all material and energy inputs and outputs are as inherently safe and benign as possible.

(5) Minimize depletion of natural resources.

(6) Strive to prevent waste.

(7) Develop and apply engineering solutions, while being cognizant of local geography, aspirations, and cultures.

(8) Create engineering solutions beyond current or dominant technologies; improve, innovate, and invent (technologies) to achieve sustainability.

(9) Actively engage communities and stakeholders in development of engineering solutions.

Chapter 5 Petroleum and Petrochemical Processing（石油加工与石化过程）

Find more information at HuaNeng, a public wechat account.

Part A Comprehensive Reading

Composition and Properties of Crude Oils

The crude oil mixture is composed of the following groups:

(1) Hydrocarbon compounds (compounds made of carbon and hydrogen).

(2) Non-hydrocarbon compounds.

(3) Organometallic compounds and inorganic salts (metallic compounds).

1 Hydrocarbon Compounds

The principal **constituents** of most crude oils are **hydrocarbon** compounds. All hydrocarbon classes are present in the crude mixture, except alkenes and alkynes. This may indicate that crude oils originated under a reducing atmosphere. The following is a brief description of the different hydrocarbon classes found in all crude oils.

Alkanes (paraffins). Alkanes are saturated hydrocarbons having the general formula C_nH_{2n+2}. The simplest alkane, methane (CH_4), is the principal constituent of natural gas. Methane, ethane, propane, and butane are gaseous hydrocarbons at ambient temperatures and atmospheric pressure. They are usually found associated with crude oils in a dissolved state.

Normal alkanes (*n*-alkanes, *n*-paraffins) are straight-chain hydrocarbons having no branches. Branched alkanes are saturated hydrocarbons with an alkyl substituent or a side branch from the main chain. A branched alkane with the same number of carbons and hydrogens as an n-alkane is called an **isomer**. For example, butane (C_4H_{10}) has two isomers, *n*-butane and 2-methyl propane (isobutane). As the molecular weight of the hydrocarbon increases, the number of isomers also increases. Pentane (C_5C_{12}) has three isomers; hexane (C_6H_{14}) has five. The following shows the isomers of hexane:

An isoparaffin is an isomer having a methyl group branching from carbon number 2 of the main chain. Crude oils contain many short, medium, and long-chain normal and branched paraffins. A naphtha fraction (obtained as a light liquid stream from crude fractionation) with a narrow boiling range may contain a limited but still large number of isomers.

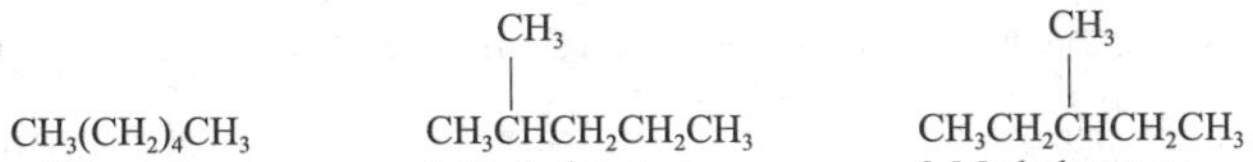

n-Hexane　2-Methylpentane (Isohexane)　3-Methylpentane

2,2-Dimethylbutane　2,3-Dimethylbutane

Cycloparaffins (Naphthenes). Saturated cyclic hydrocarbons, normally known as naphthenes, are also part of the hydrocarbon constituents of crude oils. Their ratio, however, depends on the crude type. The lower members of naphthenes are cyclopentane, cyclohexane, and their mono-substituted compounds. They are normally present in the light and the heavy naphtha fractions. Cyclohexanes, substituted cyclopentanes, and substituted cyclohexanes are important precursors for aromatic hydrocarbons.

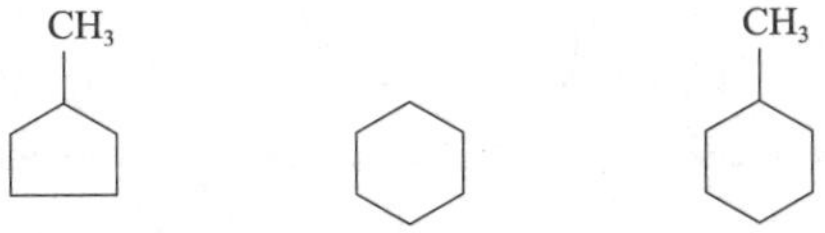

Methylcyclopentane　Cyclohexane　Methylcyclohexane

The examples shown here are for three naphthenes of special importance. If a naphtha fraction contains these compounds, the first two can be converted to benzene, and the last compound can dehydrogenate to toluene during processing. Dimethylcyclohexanes are also important precursors for xylenes. Heavier petroleum fractions such as kerosine and gas oil may contain two or more cyclohexane rings fused through two vicinal carbons.

Aromatic Compounds. Lower members of aromatic compounds are present in small amounts in crude oils and light petroleum fractions. The simplest mononuclear aromatic compound is benzene (C_6H_6). Toluene (C_7H_8) and xylene (C_8H_{10}) are also mononuclear aromatic compounds found in variable amounts in crude oils. Benzene, toluene, and xylenes (BTX) are important petrochemical intermediates as well as valuable gasoline components. Separating BTX aromatics from crude oil distillates is not feasible because they are present in low concentrations. Enriching a naphtha fraction with these aromatics is possible through a catalytic reforming process.

Binuclear aromatic hydrocarbons are found in heavier fractions than naphtha. Trinuclear and polynuclear aromatic hydrocarbons, in combination with **heterocyclic** compounds, are major constituents of heavy crudes and crude residues. Asphaltenes are a complex mixture of aromatic and heterocyclic compounds. The nature and structure of some of these compounds have been investigated. The following are representative examples of some aromatic compounds found in crude oils:

Benzene

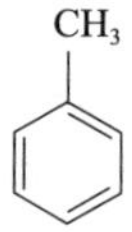

Toluene

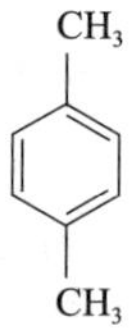

p-Xylene

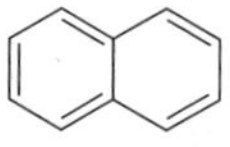

Naphthalene

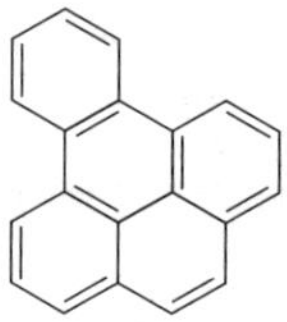

1,2-Benzopyrene

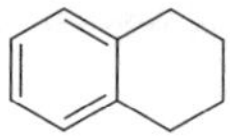

Tetralin

Only a few aromatic-cycloparaffin compounds have been isolated and identified. Tetralin is an example of this class.

2 Non–hydrocarbon Compounds

Various types of non-hydrocarbon compounds occur in crude oils and refinery streams. The most important are the organic sulfur, nitrogen, and oxygen compounds. Traces of metallic compounds are also found in all crudes. The presence of these impurities is harmful and may cause problems to certain catalytic processes. Fuels having high sulfur and nitrogen levels cause pollution problems in addition to the corrosive nature of their oxidization products.

Sulfur Compounds. Sulfur in crude oils is mainly present in the form of organosulfur compounds. Hydrogen sulfide is the only important inorganic sulfur compound found in crude oil. Its presence, however, is harmful because of its corrosive nature. Organosulfur compounds may generally be classified as acidic and non-acidic. Acidic sulfur compounds are the thiols (mercaptans). Thiophene, sulfides, and disulfides are examples of non-acidic sulfur compounds found in crude fractions. Extensive research has been carried out to identify some sulfur compounds in a narrow light petroleum fraction. Examples of some sulfur compounds from the two types are:

Acidic Sulfur Compounds

CH_3SH

Methyl mercaptan

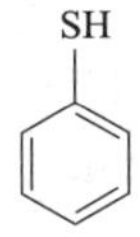

Phenyl mercaptan

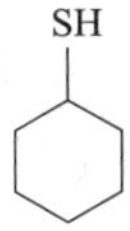

Cyclohexylthiol

Non-acidic Sulfur Compounds

CH_3SCH_3

Dimethyl sulfide

$CH_3S\text{-}SCH_3$

Dimethyldisulfide

Thiocyclohexane

Thiophene

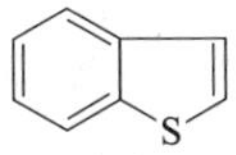

Benzothiophene

Sour crudes contain a high percentage of hydrogen sulfide. Because many organic sulfur compounds are not thermally stable, hydrogen sulfide is often produced during crude processing. High-sulfur crudes are less desirable because treating the different refinery streams for acidic hydrogen sulfide increases production costs.

Most sulfur compounds can be removed from petroleum streams through **hydrotreatment** processes, where hydrogen sulfide is produced and the corresponding hydrocarbon released. Hydrogen sulfide is then absorbed in a suitable absorbent and recovered as sulfur.

Nitrogen Compounds. Organic nitrogen compounds occur in crude oils either in a simple heterocyclic form as in pyridine (C_5H_5N) and pyrrole (C_4H_5N), or in a complex structure as in porphyrin. The nitrogen content in most crudes is very low and does not exceed 0.1wt%. In some heavy crudes, however, the nitrogen content may reach up to

0.9wt%. Nitrogen compounds are more thermally stable than sulfur compounds and accordingly are concentrated in heavier petroleum fractions and residues. Light petroleum streams may contain trace amounts of nitrogen compounds, which should be removed because they poison many processing catalysts. During hydrotreatment of petroleum fractions, nitrogen compounds are hydrodenitrogenated to ammonia and the corresponding hydrocarbon. For example, pyridine is denitrogenated to ammonia and pentane:

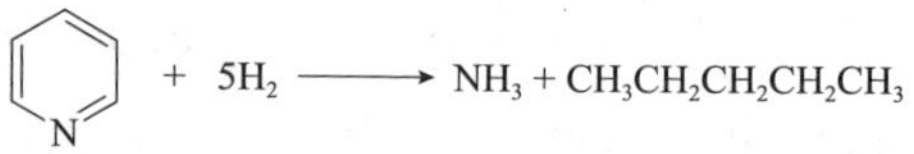

Nitrogen compounds in crudes may generally be classified into basic and non-basic categories. Basic nitrogen compounds are mainly those having a pyridine ring, and the non-basic compounds have a pyrrole structure. Both pyridine and pyrrole are stable compounds due to their aromatic nature. The following are examples of organic nitrogen compounds.

Porphyrins are non-basic nitrogen compounds. The porphyrin ring system is composed of four pyrrole rings joined by=CH-groups. The entire ring system is aromatic. Many metal ions can replace the pyrrole hydrogens and form chelates. The chelate is planar around the metal ion and resonance results in four equivalent bonds from the nitrogen atoms to the metal. Almost all crude oils and **bitumens** contain detectable amounts of vanadyl and nickel porphyrins.

Basic Nitrogen Compounds

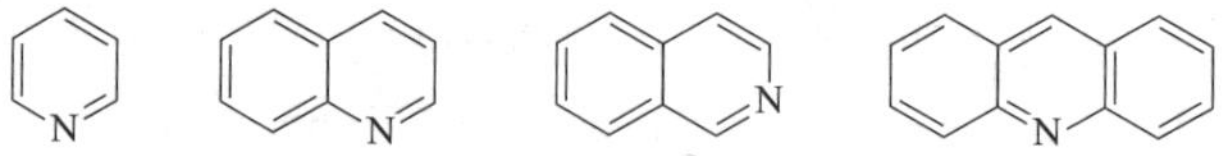

Pyridine Quinoline Isoquinoline Acridine

Non-Basic Nitrogen Compounds

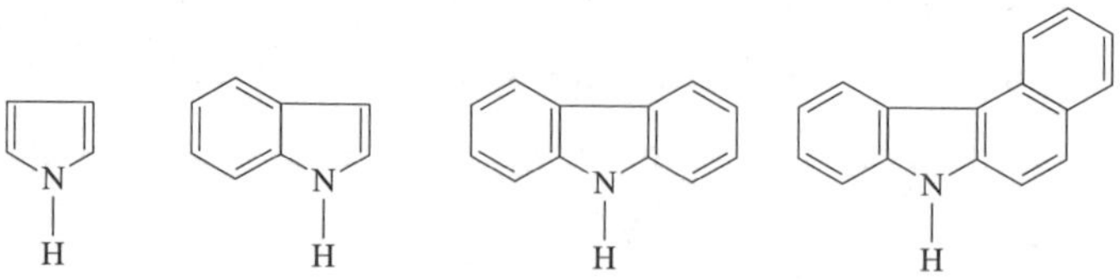

Pyrrole Indole Carbazole Benzocarbazole

Separation of nitrogen compounds is difficult, and the compounds are susceptible to alteration and loss during handling. However, the basic low-molecular weight compounds may be extracted with dilute mineral acids.

Oxygen Compounds. Oxygen compounds in crude oils are more complex than the sulfur types. However, their presence in petroleum streams is not poisonous to processing catalysts. Many of the oxygen compounds found in crude oils are weakly acidic. They are carboxylic acids, cresylic acid, phenol, and naphthenic acid. Naphthenic acids are mainly cyclopentane and cyclohexane derivatives having a carboxyalkyl side chain.

Naphthenic acids in the naphtha fraction have a special commercial importance and can be extracted by using dilute caustic solutions. The total acid content of most crudes is generally low, but may reach as much as 3%, as in some California crudes.

Non-acidic oxygen compounds such as esters, ketones, and amides are less abundant than acidic compounds. They are of no commercial value. The following shows some of the oxygen compounds commonly found in crude oils:

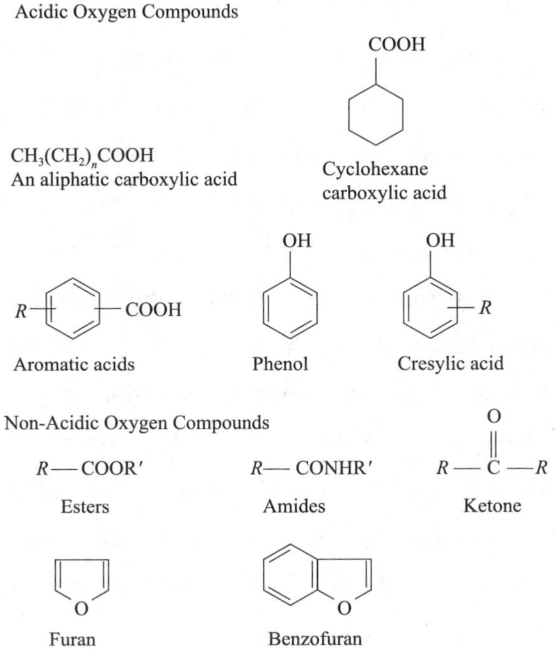

3 Metallic Compounds

Many metals occur in crude oils. Some of the more abundant are sodium, calcium, magnesium, aluminium, iron, vanadium, and nickel. They are present either as inorganic salts, such as sodium and magnesium chlorides, or in the form of organometallic compounds, such as those of nickel and vanadium (as in porphyrins). Calcium and magnesium can form salts or soaps with carboxylic acids. These compounds act as **emulsifiers**, and their presence is undesirable.

Although metals in crudes are found in trace amounts, their presence is harmful and should be removed. When crude oil is processed, sodium and magnesium chlorides produce hydrochloric acid, which is very corrosive. Desalting crude oils is a necessary step to reduce these salts.

Vanadium and nickel are poisons to many catalysts and should be reduced to very low levels. Most of the vanadium and nickel compounds are concentrated in the heavy residues. Solvent extraction processes are used to reduce the concentration of heavy metals in petroleum residues.

4 Properties of Crude Oils

Crude oils differ appreciably in their properties according to origin and the ratio of the different components in the mixture. Lighter crudes generally yield more valuable light and middle distillates and are sold at higher prices. Crudes containing a high percent of impurities, such as sulfur compounds, are less desirable than low-sulfur crudes because of their corrosivity and the extra treating cost. Corrosivity of crude oils is a function of many parameters among which are the type of sulfur compounds and their decomposition temperatures, the total acid number, the type of carboxylic and naphthenic acids in the crude and their decomposition temperatures. It was found that naphthenic acids begin to decompose at 600°F. Refinery experience has shown that above 750°F there is no naphthenic acid corrosion. For a refiner, it is necessary to establish certain criteria to relate one crude to another to be able to assess crude quality and choose the best processing scheme. The following are some of the important tests used to determine the properties of crude oils.

Density, Specific Gravity and API Gravity. Density is defined as the mass of unit volume of a material at a specific temperature. A more useful unit used by the petroleum industry is specific gravity, which is the ratio of the weight of a given volume of a material to the weight of the same volume of water measured at the same temperature.

Specific gravity is used to calculate the mass of crude oils and its products. Usually, crude oils and their liquid products are first measured on a volume basis, then changed to the corresponding masses using the specific gravity.

The API (American Petroleum Institute) gravity is another way to express the relative masses of crude oils. The API gravity could be calculated mathematically using the following equation:

$$API = \frac{141.5}{Sp.gr.60/60°} - 131.5$$

A low API gravity indicates a heavier crude oil or a petroleum product, while a higher API gravity means a lighter crude or product. Specific gravities of crude oils roughly range from 0.82 for lighter crudes to over 1.0 for heavier crudes (41~10°API scale).

Salt Content. The salt content expressed in milligrams of sodium chloride per liter oil (or in pounds/barrel) indicates the amount of salt dissolved in water. Water in crudes is mainly present in an emulsified form. A high salt content in a crude oil presents serious corrosion problems during the refining process. In addition, high salt content is a major cause of plugging heat exchangers and heater pipes. A salt content higher than 10 lb/1000 barrels (expressed as NaCl) requires desalting.

Sulfur Content. Determining the sulfur content in crudes is important because the amount of sulfur indicates the type of treatment required for the distillates. To determine sulfur content, a weighed crude sample (or fraction) is burned in an air stream. All sulfur compounds are oxidized to sulfur dioxide, which is further oxidized to sulfur trioxide and finally titrated with a standard alkali.

Identifying sulfur compounds in crude oils and their products is of little use to a refiner because all sulfur compounds can easily be hydrodesulfurized to hydrogen sulfide and the corresponding hydrocarbon. The sulfur content of crudes, however, is important and is usually considered when determining commercial values.

Pour Point. The pour point of a crude oil or product is the lowest temperature at which an oil is observed to flow under the conditions of the test. Pour point data indicates the amount of long-chain paraffins (petroleum wax) found in a crude oil. Paraffinic crudes usually have higher wax content than other crude types. Handling and transporting crude oils and heavy fuels is difficult at temperatures below their pour points Often, chemical additives known as pour point depressants are used to improve the flow properties of the fuel. Long-chain n-paraffins ranging from 16~60 carbon atoms in particular, are responsible for near-ambient temperature precipitation. In middle distillates, less than 1% wax can be sufficient to cause solidification of the fuel.

Ash Content. This test indicates the amount of metallic constituents in a crude oil. The ash left after completely burning an oil sample usually consists of stable metallic salts, metal oxides, and silicon oxide. The ash could be further analyzed for individual elements using spectroscopic techniques.

Selected from Chapter 1 of "*Chemistry of Petrochemical Processes* (2nd edition)", Sami Matar, Gulf Publishing Company, 2000.

Notes

Constituent: *adjective.* (1) being a part of a whole. (2) being a voting member of a community or organization and having the power to appoint or elect. *noun.* (1) a member of a constituency. (2) a component part of something.

Hydrocarbon: *noun.* a compound of hydrogen and carbon, such as any of those which are the chief components of petroleum and natural gas.

Isomer: *noun.* (1) each of two or more compounds with the same formula but a different arrangement of atoms in the molecule and different properties. (2) each of two or more atomic nuclei that have the same atomic number and the same mass number but different energy states.

Heterocyclic: *adjective.* denoting a compound whose molecule contains a ring of atoms of at least two elements (one of which is generally carbon).

Hydrotreatment: Hydrotreating processes aim at the removal of impurities such as sulfur and nitrogen from distillate fuels—naphtha, kerosene, and diesel—by treating the feed with hydrogen at elevated temperature and pressure in the presence of a catalyst.

Bitumen: *noun.* a black viscous mixture of hydrocarbons obtained naturally or as a residue from petroleum distillation. It is used for road surfacing and roofing.

Emulsifier: *noun.* a substance that stabilizes an emulsion, in particular a food additive used to stabilize processed foods.

Questions

(1) Biomass and crude oil can be converted to chemicals by different reactions due to the difference of the compositions. What are the compositions of crude oils and biomass, respectively?

(2) What is the definition of isomer? Please try to list all the isomers of hexane.

(3) Saturated cyclic hydrocarbons, normally known as naphthenes, are important precursors for aromatic hydrocarbons. Please check the reference and provide some details for converting cyclohexane to benzene.

(4) Crude oils contain a small amount of aromatic compounds. Aromatic compounds, especially benzene, toluene, and xylenes, are important petrochemical intermediates and valuable gasoline components. Please check the literature to illustrate the important of aromatic compounds for industrial application.

(5) Why most sulfur compounds should be removed from petroleum streams? What processes can be used to remove sulfur from petroleum streams?

(6) Organic nitrogen compounds occur in crude. Does nitrogen compound need to be removed from the petroleum streams? Please explain the reason and provide an example.

(7) Oxygen compounds occur in crude oil. Many of the oxygen compound are weakly acidic. Please give some examples of acidic oxygen–containing components in crude oil.

(8) Crude oil contains many kinds of metal compounds. For example, sodium, magnesium, vanadium and nickel. Does the metal compounds need to be removed from the crude oil? Please explain.

(9) Crude oil contains many impurities. The property of crude oil is a function of many parameters. Thus, it is necessary to establish certain criteria to assess crude oil quality and choose the best processing scheme. Please list and explain some important tests and criteria to determine the properties of crude oils.

(10) Have you heard about crude oil to chemicals? Do you know the chemistry and process for such novel technology?

Part B Reading for Innovation

Crude Oil Processing and Production of Hydrocarbon Intermediates

1 Physical Separation Processes

Physical separation techniques separate a mixture such as a crude oil without changing the chemical characteristics of the components. The separation is based on differences of certain physical properties of the constituents such as the boiling and melting points, adsorption affinities on a certain solid, and diffusion through certain membranes. The important physical separation processes, discussed here, are distillation, absorption, adsorption, and solvent extraction.

(1) Atmospheric Distillation

Atmospheric **distillation** separates the crude oil complex mixture into different fractions with relatively narrow boiling ranges. In general, separation of a mixture into fractions is based primarily on the difference in the boiling points of the components. In atmospheric distillation units, one or more **fractionating** columns are used.

Distilling a crude oil starts by preheating the feed by exchange with the hot product streams. The feed is further heated to about 320℃ as it passes through the heater pipe (pipe still heater).

The hot feed enters the fractionator, which normally contains 30~50 fractionation trays. Steam is introduced at the bottom of the fractionator to strip off light components. The efficiency of separation is a function of the number of theoretical plates of the fractionating tower and the reflux ratio. Reflux is provided by condensing part of the tower overhead vapors. Reflux ratio is the ratio of vapors condensing back to the still to vapors condensing out of the still (distillate). The higher the reflux ratio, the better the separation of the mixture.

Products are withdrawn from the distillation tower as side streams, while the reflux is provided by returning a portion of the cooled vapors from the tower overhead condenser. Additional reflux could be obtained by returning part of the cold side stream products to the tower. In practice, the **reflux** ratio varies over a wide range according to the specific separations desired. From the **overhead** condenser, the uncondensed gases are separated, and the condensed light naphtha liquid is withdrawn to storage. Heavy naphtha, kerosine, and gas oil are withdrawn as side stream products. Table 1 shows the approximate boiling ranges for crude oil fractions. The residue (topped crude) is removed from the bottom of the distillation tower and may be used as a fuel oil. It may also be charged to a vacuum distillation unit, a catalytic cracking or steam cracking process. Figure 1 is a flow diagram for atmospheric and vacuum distillation units.

Table 1. Approximate ASTM boiling point ranges for crude oil fractions

	Boiling range	
Fractions	℉	℃
Light naphtha	85~210	30~99
Heavy naphtha	190~400	88~204
Kerosine	340~520	171~271
Atmospheric gas oil	540~820	288~438
Vacuum gas oil	750~1050	399~566
Vacuum residue	1000+	538+

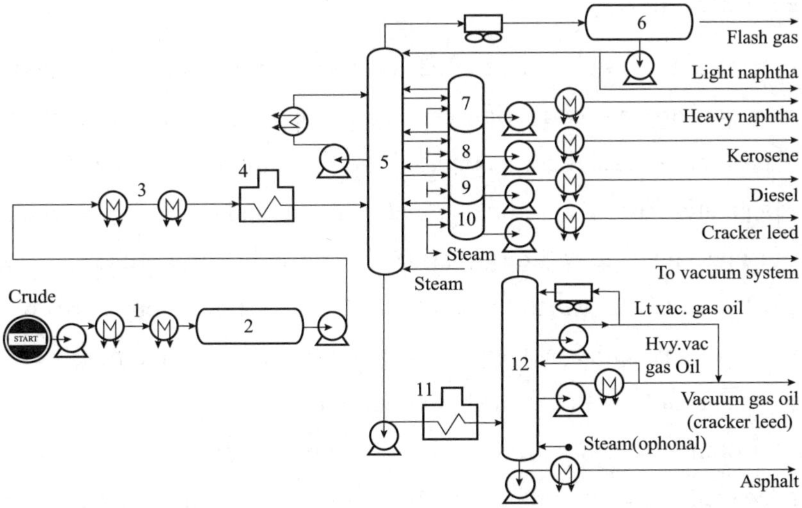

Figure 1 Flow diagram of atmospheric and vacuum distillation units: (1,3) heat exchangers; (2) desalter, (3,4) heater; (5) distillation column, (6) overhead condenser, (7~10) pump around streams, (11) vacuum distillation heater; (12) vacuum tower

(2) Vacuum Distillation

Vacuum distillation increases the amount of the middle distillates and produces lubricating oil base stocks and asphalt. The feed to the unit is the residue from atmospheric distillation. In vacuum distillation, reduced pressures are applied to avoid cracking long-chain hydrocarbons present in the feed.

The feed is first preheated by exchange with the products, charged to the vacuum unit heater, and then passed to the vacuum tower in an atmosphere of superheated steam. Using superheated steam is important: it decreases the partial pressure of the hydrocarbons and reduces coke formation in the furnace tubes. Distillation normally occurs at a temperature range of 400~440°C and an absolute pressure of 25~40mmHg. The top tower temperature is adjusted by refluxing part of the gas oil product (top product). The size (diameter) of the vacuum distillation tower is much larger than atmospheric towers because the volume of the vapor/unit-volume of the feed is much larger than in atmospheric distillation.

Products obtained as side streams are vacuum gas oil (VGO), lube oil base stocks, and asphalt. Asphalt may be used for paving roads or may be charged to a delayed coking unit.

(3) Solvent Extraction

Liquid solvents are used to extract either desirable or undesirable compounds from a liquid mixture. Solvent extraction processes use a liquid solvent that has a high solvolytic power for certain compounds in the feed mixture. For example, ethylene glycol has a greater affinity for aromatic hydrocarbons and extracts them preferentially from a reformate mixture (a liquid paraffinic and aromatic product from catalytic reforming). The raffinate, which is mainly paraffins, is freed from traces of ethylene glycol by distillation. Other solvents that could be used for this purpose are liquid sulfur dioxide and sulfolane (tetramethylene sulfone).

The sulfolane process is a versatile extractant for producing high purity BTX aromatics (benzene, toluene, and xylenes). It also extracts aromatics from kerosines to produce low-aromatic jet fuels.

On the other hand, liquid propane also has a high affinity for paraffinic hydrocarbons. Propane deasphalting removes asphaltic materials from heavy lube oil base stocks. These materials reduce the viscosity index of lube oils. In this process, liquid propane dissolves mainly paraffinic hydrocarbons and leaves out asphaltic materials. Higher extraction temperatures favor better separation of the asphaltic components. Deasphalted oil is stripped to recover propane, which is recycled.

Solvent extraction may also be used to reduce asphaltenes and metals from heavy fractions and residues before using them in catalytic cracking. The organic solvent separates the resids into demetallized oil with lower metal and asphaltene content than the feed, and asphalt with high metal content. Figure 2 shows the IFP deasphalting process.

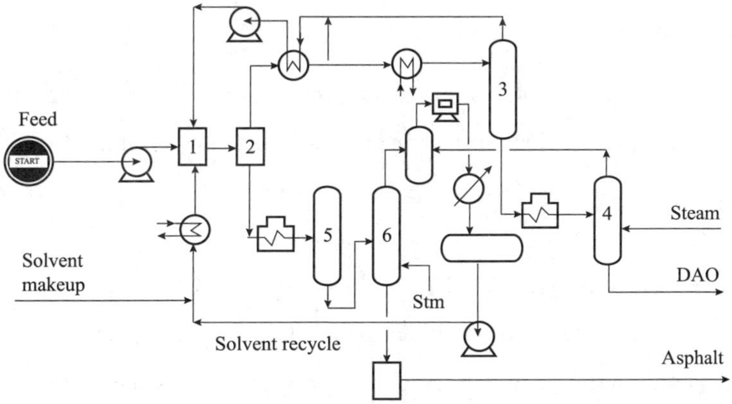

Figure 2 The IFP deasphalting process:4 (1,2) extractor, (3~6) solvent recovery towers

Solvent extraction is used extensively in the petroleum refining industry. Each process uses its selective solvent, but, the basic principle is the same as above.

2 Thermal Conversion Processes

Thermal cracking was the first process used to increase gasoline production. After the development of catalytic cracking, which improved yields and product quality, thermal cracking was given other roles in refinery operations. The three important thermal cracking techniques are coking, viscosity breaking, and steam cracking. Steam cracking is of special importance as a major process designed specifically for producing light olefins.

(1) Coking Processes

Coking is a severe thermal cracking process designed to handle heavy residues with high asphaltene and metal contents. These residues cannot be fed to catalytic cracking units because their impurities deactivate and poison the catalysts.

Products from coking processes vary considerably with feed type and process conditions. These products are hydrocarbon gases, cracked naphtha, middle distillates, and coke. The gas and liquid products are characterized by a high percentage of unsaturation. Hydrotreatment is usually required to saturate olefinic compounds and to desulfurize products from coking units.

(2) Thermal Cracking Reactions

The first step in cracking is the thermal decomposition of hydrocarbon molecules to two free radical fragments. This initiation step can occur by a homolytic carbon-carbon bond scission at any position along the hydrocarbon chain. The following represents the initiation reaction:

$$RCH_2CH_2CH_2R' \longrightarrow RCH_2\dot{C}H_2 + R'\dot{C}H_2$$

The radicals may further crack, yielding an olefin and a new free radical. Cracking usually occurs at a bond beta to the carbon carrying the unpaired electron.

$$RCH_2\dot{C}H_2 \longrightarrow \dot{R} + CH_2{=\!=}CH_2$$

Free radicals may also react with a hydrocarbon molecule from the feed by abstracting a hydrogen atom. In this case the attacking radical is **terminated**, and a new free radical is formed. Abstraction of a hydrogen atom can occur at any position along the chain. However, the rate of hydrogen abstraction is faster from a tertiary position than from a secondary, which is faster than from a primary position.

3 Catalytic Reforming

Catalytic reforming is the process of transforming C_7~C_{10} hydrocarbons with low octane numbers to aromatics and iso-paraffins, which have high octane numbers. It is a highly endothermic process requiring large amounts of energy. A schematic presentation of the feedstock, products and process condition is shown in Figure 3. The process can be operated in two modes: a high severity mode to produce mainly aromatics (80vol%~90vol%) and a middle severity mode to produce high octane gasoline (70% aromatics content).

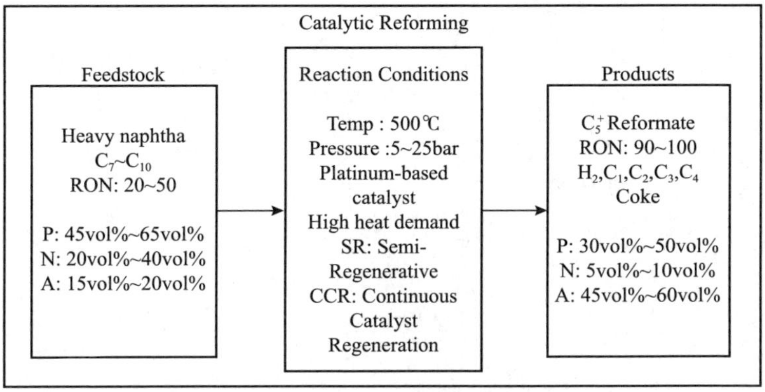

Figure 3 Catalytic reforming process

The catalytic reformer is one of the major units for gasoline production in refineries. It can produce 37wt% of the total gasoline pool. Other units such as the fluid catalytic cracker (FCC), the methyl ter-butyl ether (MTBE) production unit, alkylation unit and isomerization unit, also contribute to this pool.

The straight run naphtha from the crude distillation unit is hydrotreated to remove sulphur, nitrogen and oxygen which can all deactivate the reforming catalyst. The hydrotreated naphtha (HTN) is fractionated into light naphtha (LN), which is mainly C_5~C_6, and heavy naphtha (HN) which is mainly C_7~C_{10} hydrocarbons. It is important to remove C_6 from the reformer feed because it will form benzene which is considered carcinogenic upon combustion. Light naphtha (LN) is isomerized in the isomerization unit (I). Light naphtha can be cracked if introduced to the reformer. The role of the heavy naphtha (HN) reformer in the refinery is shown in Figure 4. Hydrogen, produced in the reformer can be recycled to the naphtha hydrotreater, and the rest is sent to other units demanding hydrogen.

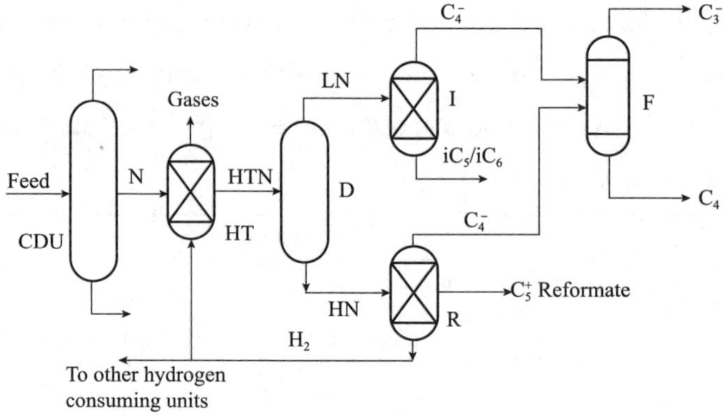

CDU	Crude distillation unit
D	Distillation
F	Flash
I	Isomerization
HN	Heavy Naphtha
HT	Hydrotreater
HTN	Hydro Treated Naphtha
LN	Light Naphtha
N	Naphtha
R	Reformer

Figure 4 Role of reformer in the refinery

The following reaction occur during reforming:

Naphthene Dehydrogenation of Cyclohexanes

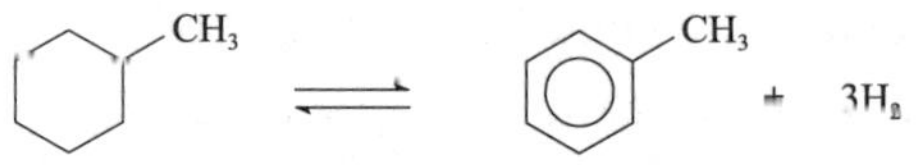

Methyl Cyclohexane Toluene

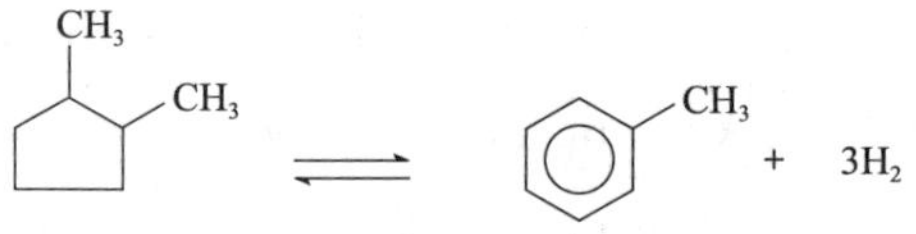

Dimethyl-Cyclopentane Toluene

Paraffin Dehydrogenation

$$n\text{-}C_7H_{16} \rightleftharpoons n\text{-}C_7H_{14} + H_2$$

n-heptane *n*-heptene

Dehydrocyclization

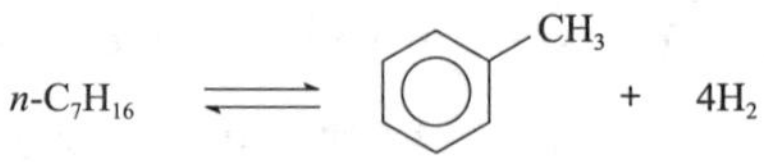

n-heptane Toluene

Isomerization

$$n\text{-}C_7H_{16} \rightleftharpoons iC_7H_{16}$$

n-heptane isoheptane

Paraffin hydrocracking

$$C_{10}H_{22} + H_2 \longrightarrow C_2H_5-\underset{\displaystyle CH_3}{\underset{|}{CH}}-C_2H_5 + C_4H_{10}$$

Decane Isohexane

Hydrocracking of aromatics

Semi–regenerative Fixed Bed Process. The schematic flow diagram of this process is shown in Figure 5. The name semi-regenerative comes from **regeneration** of the catalyst in the fixed bed reactors after shut down by burning off the carbon formed on the catalyst surface.

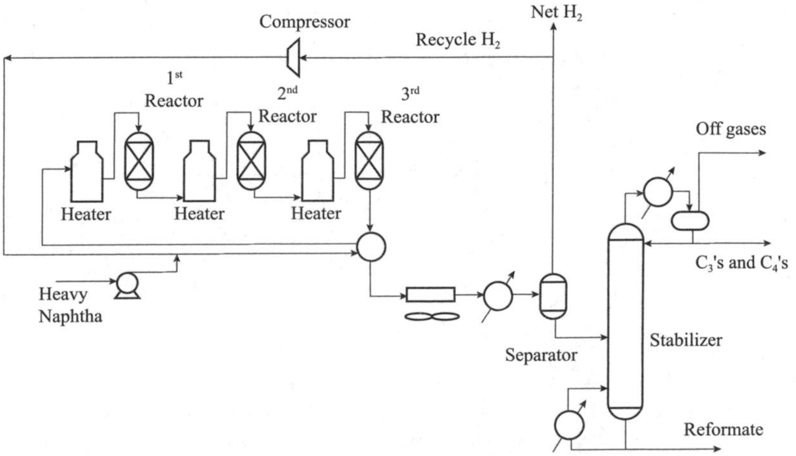

Figure 5 Semi–regenerative (SR) fixed bed reforming process

Reactions such as dehydrogenation of paraffins and naphthenes which are very rapid and highly endothermic occur in the first reactor, with high temperature drop. Reactions that are considered rapid, such as paraffin isomerization and naphthens dehydroisomerization, give moderate temperature decline in the second reactor. Furthermore, slow reactions such as dehydrocyclization and hydrocracking give low temperature decline in the third reactor.

The temperature and concentration profile in each reactor is shown in Figure 6. To prevent catalyst coking, the hydrogen partial pressure is maintained at a level such that the hydrogen-to-hydrocarbon ratio by weight (H_2/HC) is greater than 25 for monometallic catalyst. This is done by recycling some of the hydrogen produced (Figure 5). Some light hydrocarbons (C_1~C_4) are separated from the reformate in the stabilizer. At the top of the stabilizer residual hydrogen and C_1 to C_4 are withdrawn as condenser products, which are then sent to gas processing, and part of the liquid product (C_3 and C_4) is returned from the reflux drum back to the stabilizer. The main product of the column is stabilized reformate, which is sent to the gasoline blending plant.

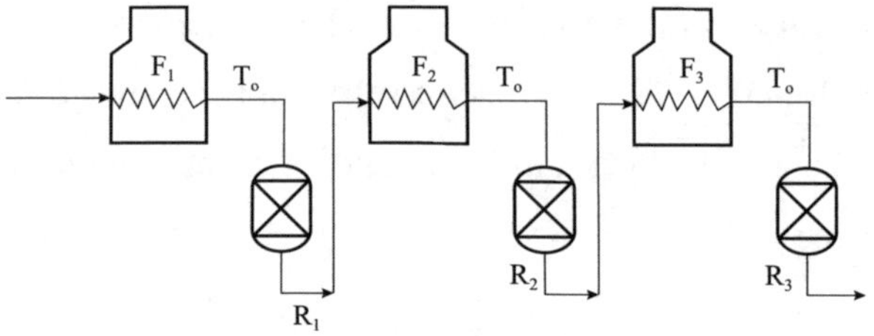

(A)Furnace and reactor layout for fixed bed reformer

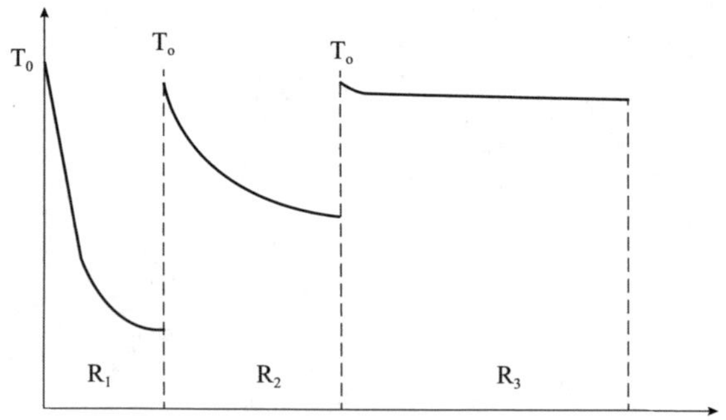

(B)Variation of temperature in the reactors

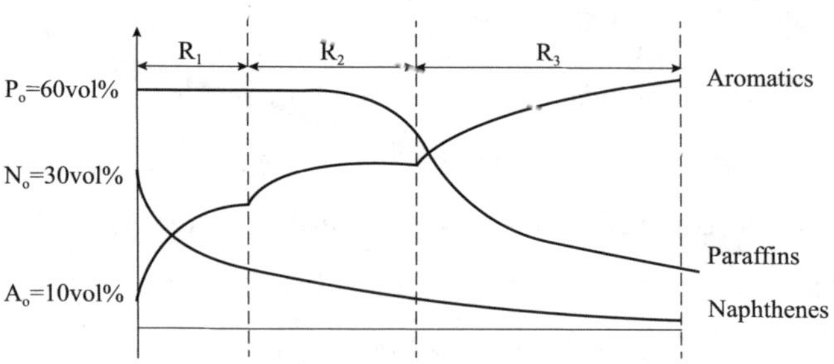

(C) Variation in effluent compositions

Figure 6 P_o, initial Paraffins; N_o, initial Napthenes and A_o, initial Aromatics

A slight modification to the semi-regenerative process is to add an extra reactor to avoid shutting down the whole unit during regeneration. Three reactors can be running while the forth is being regenerated. This modified process is called the “cyclic fixed bed” process.

Continuous Regenerative (moving bed) CCR Platforming UOP Process. In this process, three or four reactors are installed one on the top of the other. The schematic flow diagram of the continuous regenerative process (CCR) is shown in Figure 7. UOP has licensed this process under the CCR Platforming process. The effluent from each reactor is sent to a common furnace for heating. The catalyst moves downwards by gravity from the first reactor (R_1) to the forth reactor (R_4). The catalyst is sent to the regenerator to burn off the coke and then sent back to the first reactor R_1. The final product from R_4 is sent to the stabilizer and gas recovery section. The process can be operated at lower hydrogen partial pressure (P_{H_2}=3bar) compared to the semi-generative process (P_{H_2}=35bar), with a reformate yield gain of nearly 10vol%.

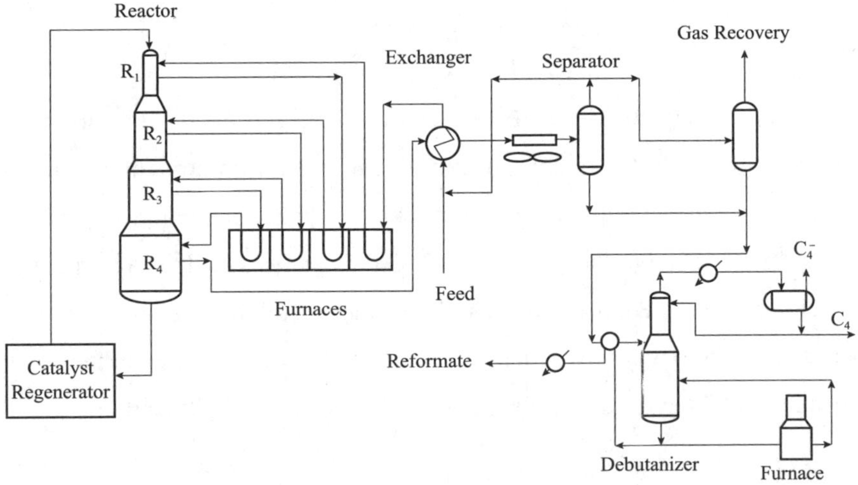

Figure 7 Continuous regenerative reformer (CCR), UOP Platforming process

4 Catalytic Cracking

Catalytic cracking (Cat-cracking) is a remarkably versatile and flexible process. Its principal aim is to crack lower-value stocks and produce higher-value light and middle distillates. The process also produces light hydrocarbon gases, which are important feedstocks for petrochemicals. Catalytic cracking produces more gasoline of higher octane than thermal cracking. This is due to the effect of the catalyst, which promotes isomerization and dehydrocyclization reactions.

Catalyst. Acid-treated clays were the first catalysts used in catalytic cracking processes, but have been replaced by synthetic amorphous silica-alumina, which is more active and stable. Incorporating zeolites (crystalline alumina-silica) with the silica/alumina catalyst improves selectivity towards aromatics. These catalysts have both Lewis and Bronsted acid sites that promote **carbonium** ion formation. An important structural feature of zeolites is the presence of holes in the crystal lattice, which are formed by the silica-alumina tetrahedra. Each tetrahedron is made of four oxygen anions with either an aluminum or a silicon cation in the center. Each oxygen anion with a -2 oxidation state is shared between either two silicon, two aluminum, or an aluminum and a silicon cation.

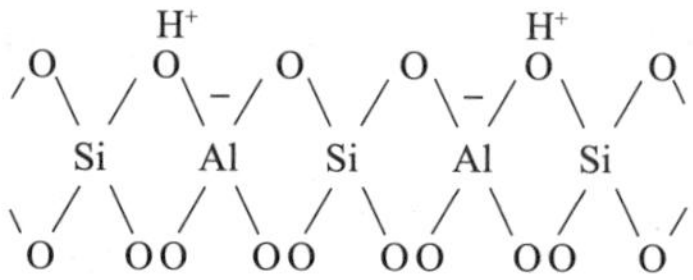

Reaction. A major difference between thermal and catalytic cracking is that reactions through catalytic cracking occur via carbocation intermediate, compared to the free radical intermediate in thermal cracking. Carbocations are longer lived and accordingly more selective than free radicals. Acid catalysts such as amorphous silica-alumina and crystalline zeolites promote the formation of carbocations. The following illustrates the different ways by which carbocations may be generated in the reactor:

Abstraction of a hydride ion by a Lewis acid site from a hydrocarbon

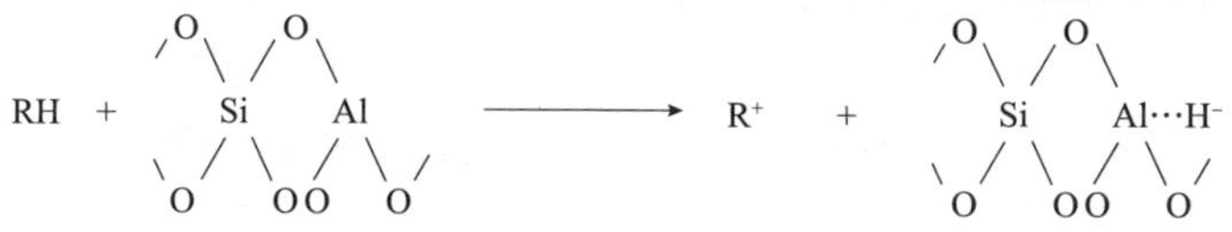

Lewis Acid Site

Reaction between a Bronsted acid site (H^+) and an olefin

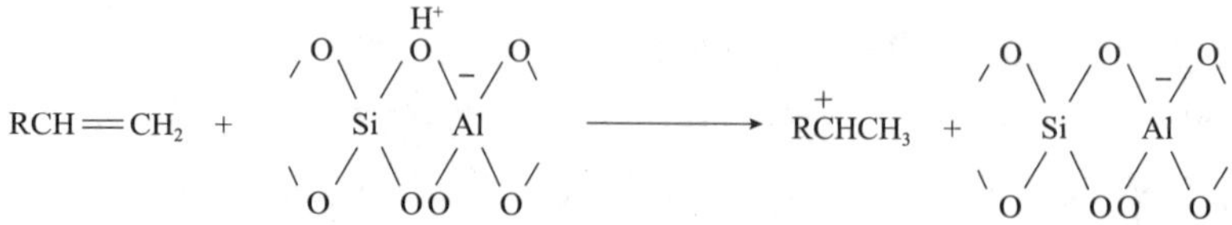

Reaction of a carbonium ion formed from step 1 or 2 with another hydrocarbon by abstraction of a hydride ion

$$R^+ + RCH_2CH_3 \longrightarrow RH + R\overset{+}{C}HCH_3$$

Abstraction of a hydride ion from a tertiary carbon is easier than from a secondary, which is easier than from a primary position. The formed carbocation can rearrange through a methide-hydride shift similar to what has been explained in catalytic reforming. This isomerization reaction is responsible for a high ratio of branched isomers in the products. The most important cracking reaction, however, is the carbon-carbon beta bond scission. A bond at a position beta to the positively-charged carbon breaks heterolytically, yielding an olefin and another carbocation. This can be represented by the following example:

$$RCH_2\overset{+}{C}HCH_3 \longrightarrow R^+ + CH_2{=}CHCH_3$$

The carbon-carbon beta scission may occur on either side of the carbocation, with the smallest fragment usually containing at least three carbon atoms. For example, cracking a secondary carbocation formed from a long chain paraffin could be represented as follows:

$$RCH_2\overset{+}{C}HCH_2CH_2R' \xrightarrow{a} R^+ + CH_2{=}CHCH_2CH_2R'$$

$$RCH_2\overset{+}{C}HCH_2CH_2R' \xrightarrow{b} R'\overset{+}{C}H_2 + RCH_2CH{=}CH_2$$

Aromatization of paraffins can occur through a dehydrocyclization reaction. Olefinic compounds formed by the beta scission can form a carbocation intermediate with the configuration conducive to cyclization. For example, if a carbocation such as that shown below is formed (by any of the methods mentioned earlier), cyclization is likely to occur.

Once cyclization has occurred, the formed carbocation can lose a proton, and a cyclohexene derivative is obtained. This reaction is aided by the presence of an olefin in the vicinity ($R{-}CH{=}CH_2$).

$$RCH_2CH_2CH_2CH_2CH{=}CH_2 \xrightarrow{\text{zeolite}} R\overset{+}{C}HCH_2CH_2CH_2CH{=}CH_2$$

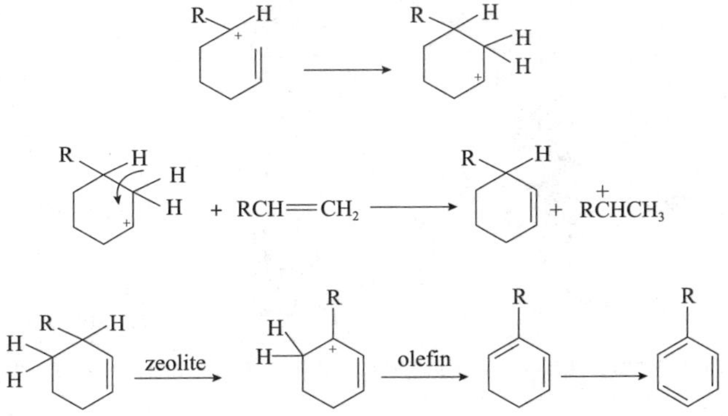

During the cracking process, fragmentation of complex polynuclear cyclic compounds may occur, leading to formation of simple cycloparaffins. These compounds can be a source of C_6, $C_7$and C_8 aromatics through isomerization and hydrogen transfer reactions.

Coke formed on the catalyst surface is thought to be due to polycondensation of aromatic nuclei. The reaction can also occur through a carbonium ion intermediate of the benzene ring. The polynuclear aromatic structure has a high C/H ratio.

Process. Fluidized catalytic cracking (FCC) is a process in which lighter boiling fractions can be generated from heavy petroleum stocks ranging from VGO to residues. Zeolite, silica or alumina acts as the cracking catalyst. The temperature of cracking reactions is in the range of 500~600°C.

Fine catalyst powder is fluidized in the tall tubular part (known as riser) of the reactor with the help of steam and light vaporizable hydrocarbons, like naphtha, which reduces the viscosity of the feed oil. Heavy feed oil is atomized by steam in the riser. Products are carried to the wider disengagement section of the reactor where products are disengaged from the catalyst through multistage cyclones. Hydrocarbon vapors from the reactor then enters a distillation column with side strippers for the recovery of various fractions.

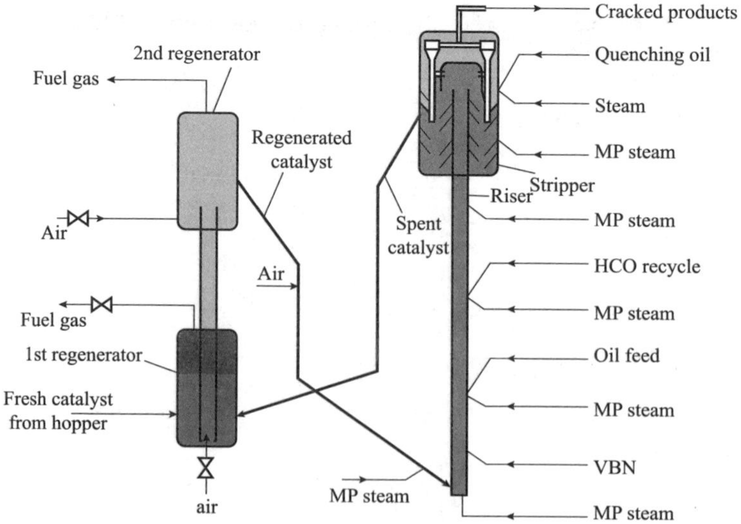

Figure 8 A modern catalytic cracking reactor unit with separate reactor and regenerator

Spent catalyst laden with coke is transported to a fluidized bed regenerator either below or at the side of the reactor where coke on the catalyst is burnt out to regenerate the catalyst from its temporary deactivation. The regenerated and hot catalyst is then returned to the riser reactor to continue the process of cracking. Air is used as the fluidization medium in the regenerator and the heat of combustion is used to maintain the reaction temperature. A small amount of coke (called equilibrium coke) is always retained on the surface and additional fresh catalyst must be maintained to compensate for the loss of activity.

The organometallic hydrocarbons in the feed deposits metals on the surface of the catalyst, causing permanent deactivation of the catalyst. Usually, vanadium, nickel, and sodium are found in the feedstock and cause the deactivation. The presence of nitrogen in the feed also destroys the acid sites of the catalyst. Gases, gasoline, light cycle oil, heavy cycle oil, and residual oil contaminated with the catalyst are obtained from the distillation column. The light cycle oil goes to the diesel pool and the heavy cycle oil goes to the fuel oil pool. Residual oil goes to a catalyst decantation unit before it is recycled to the reactor. Decanted oil also acts as a quenching medium in the reactor. Gases and gasoline from the distillation column must be treated for sulfur removal separately. Because of the presence of olefi nic hydrocarbons in the gasoline, hydrogenation may be required to improve its oxidation stability.

Selected from "*Chemistry of Petrochemical Processes* (2nd edition)", Sami Matar, Gulf Publishing Company, 2000.

Selected from "*Fundamentals of Petroleum Refining*", Taher A. Alsahhaf, Amal Sayed Elkilani, M.A. Fahim, Elsevier, 2010.

Important Words and Phrases

Distillation: *noun.* (1) the action of purifying a liquid by a process of heating and cooling. (2) the extraction of the essential meaning or most important aspects of something.

Fractionate: *verb.* divide into fractions or components.

Reflux: *noun.* (1) the flow of a fluid through a vessel or valve in the body in a direction opposite to normal. (2) the process of boiling a liquid so that any vapor is liquefied and returned to the stock. *verb.* (1) (of a fluid) flow backwards through a vessel or valve in the body. (2) boil or cause to boil in circumstances such that the vapor returns to the stock of liquid after condensing.

Overhead: *adjective.* (1) situated above the level of the head. (2) (of a driving mechanism) above the object driven. (3) (of a cost or expense) incurred in the general upkeep or running of a plant, premises, or business, and not attributable to specific products or items. *noun.* (1) an overhead cost or expense. (2) a transparency designed for use with an overhead projector. (3) an overhead compartment.

Terminate: *verb.* bring to an end.

Regeneration: *noun.* the action or process of regenerating or being regenerated, in particular the formation of new animal or plant tissue.

Carbonium: *noun.* an organic cation having one less electron than a corresponding free radical and with positive charge localized on carbon atom.

Part C Reading for Fun

Synthetic Polymers

1 Introduction

The synthetic polymer industry represents the major end use of many petrochemical monomers such as ethylene, styrene, and vinyl chloride. Synthetic polymers have greatly affected our lifestyle. Many articles that were previously made from naturally occurring materials such as wood, cotton, wool, iron, aluminum, and glass are being replaced or partially substituted by synthetic polymers. Clothes made from polyester, nylon, and acrylic fibers or their blends with natural fibers currently dominate the apparel market. Plastics are replacing many articles previously made of iron, wood, porcelain, or paper in thousands of diversified applications. Polymerization could now be tailored to synthesize materials stronger than steel. For example, polyethylene fibers with a molecular weight of one million can be treated to be 10 times stronger than steel!

However, its melting point is 148°C. A recently announced thermotropic liquid crystal polymer based on p-hydroxybenzoic acid, terephthalic acid, and p, pv-biphenol has a high melting point of 420°C and does not decompose up to 560°C. This polymer has an initial stress of 3.4×10^6kg/mm^2, even after 6000 hours of testing.

The polymer field is versatile and fast growing, and many new polymers are continually being produced or improved. The basic chemistry principles involved in polymer synthesis have not changed much since the beginning of polymer production. Major changes in the last 70 years have occurred in the catalyst field and in process development. These improvements have a great impact on the economy. In the elastomer field, for example, improvements influenced the automobile industry and also related fields such as mechanical goods and wire and cable insulation.

Most big-volume thermoplastics are produced by addition polymerization. Other thermoplastics are synthesized by condensation. Table 1 shows the major thermoplastics.

Table 1. Major thermoplastic polymers

Name	Family	Formula	Melting T /℃	Density/(g/cm³)
Low-density polyethylene	polyolefin	$(C_2H_4)_n$	85~110	0.91 ~0.93
High-density polyethylene	polyolefin	$(C_2H_4)_n$	85~110	0.94~0.96
Polypropylene	polyolefin	$(C_3H_6)_n$	189	0.89~0.91
Polyvinyl chloride	Vinyl	$(C_2H_3Cl)_n$	—	1.38
Polyvinyl acetate	Vinyl	$(C_4H_6O_2)_n$	60	—
Polystyrene	Styrenic	$(C_8H_8)_n$	212	1.05
Acrylonitrile butadiene styrene	Styrenic	$(C_8H_8)_x\cdot(C_4H_6)_y\cdot(C_3H_3N)_z$	—	1.04
Acrylonitrile styrene	Styrenic	$(C_8H_8)_x\cdot(C_3H_3N)_y$	—	1.06~1.08
Polymethylmethacrylate	Acrylic	$[CH_2C(CH_3)(CO_2CH_3)]_n$	—	1.2
Polycaprolactam	Polyamide	$C_6H_{13}NO$	220	0.9~1.0
Polyethylene-terephthalate	Polyester	$(C_{10}H_8O_4)_n$	250~255	1.68
Polybutylene-terephthalate	Polyester	$C_{12}H_{16}O_6$	227	1.31
Polycarbonates	Polyester	$(OROCO)_n$	220~230	1.2
Polyacetals	Polyethers	$(CH_2O)_n$	165~175	1.39~1.43

2 Polyethylene

Polyethylene is the most extensively used thermoplastic. The ever increasing demand for polyethylene is partly due to the availability of the monomer from abundant raw materials (associated gas, LPG, naphtha). Other factors are its relatively low cost, ease of processing the polymer, resistance to chemicals, and its flexibility.

High-pressure polymerization of ethylene was introduced in the 1930s. The discovery of a new titanium catalyst by Karl Ziegler in 1953 revolutionized the production of linear unbranched polyethylene at lower pressures. The two most widely used grades of polyethylene are low-density polyethylene (LDPE) and high-density polyethylene (HDPE). Currently, a new LDPE grade has been introduced. It is a linear, low-density grade (LLDPE) produced like the high-density polymer at low pressures. Polymerizing ethylene is a highly exothermic reaction. For each gram of ethylene consumed, approximately 3.5kJ (850cal) are released:

$$nCH_2{=}CH_2 \longrightarrow [CH_2{-}CH_2]_n \quad \Delta H = -92\text{kJ/mol}$$

Low-density polyethylene (LDPE) is produced under high pressure in the presence of a free radical initiator. As with many free radical chain addition polymerizations, the polymer is highly branched. It has a lower crystallinity compared to HDPE due to its lower capability of packing.

Polymerizing ethylene can occur either in a tubular or in a stirred autoclave reactor. In the stirred autoclave, the heat of the reaction is absorbed by the cold ethylene feed. Stirring keeps a uniform temperature throughout the reaction vessel and prevents agglomeration of the polymer.

Reaction conditions for the free radical polymerization of ethylene are 100~200°C and 100~135 atmospheres. Ethylene conversion is kept to a low level (10%~25%) to control the heat and the viscosity. However, overall conversion with recycle is over 95%.

High-density polyethylene (HDPE) is produced by a low-pressure process in a fluid-bed reactor. Catalysts used for HDPE are either of the Zieglar-type [a complex of $Al(C_2H_5)_3$ and α-$TiCl_4$] or silicaalumina impregnated with a metal oxide such as chromium oxide or molybdenum oxide.

Reaction conditions are generally mild, but they differ from one process to another. In the newer Unipol process (Figure 1) used to produce both HDPE and LLDPE, the reaction occurs in the gas phase. Ethylene and the comonomers (propene, 1-butene, etc.) are fed to the reactor containing a fluidized bed of growing polymer particles. Operation temperature and pressure are approximately 100°C and 20 atmospheres. A single-stage centrifugal compressor circulates unreacted ethylene. The circulated gas fluidizes the bed and removes some of the exothermic reaction heat. The product from the reactor is mixed with additives and then pelletized.

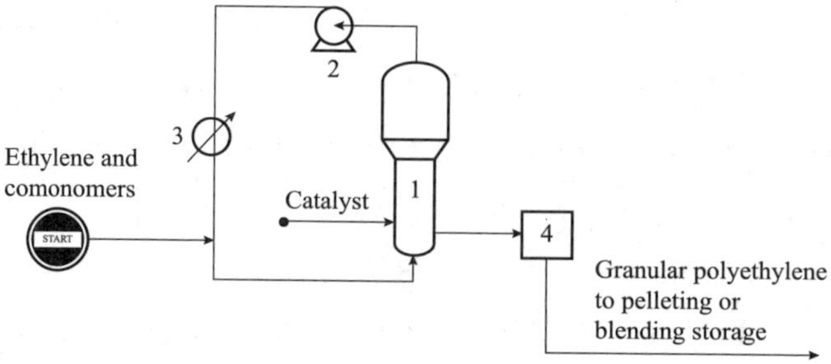

Figure 1 The Union Carbide Unipol process for producing HDPE: (1) reactor, (2) single-stage centrifugal compressor, (3) heat exchanger, (4) discharge tank.

Polyethylene is an inexpensive thermoplastic that can be molded into almost any shape, extruded into fiber or filament, and blown or precipitated into film or foil. Polyethylene products include packaging (largest market), bottles, irrigation pipes, film, sheets, and insulation materials.

3 Polyproylene

Polypropylene (PP) is a major thermoplastic polymer. Although polypropylene did not take its position among the large volume polymers until fairly recently, it is currently the third largest thermoplastic after PVC. The delay in polypropylene development may be attributed to technical reasons related to its polymerization. Polypropylene produced by free radical initiation is mainly the atactic form. Due to its low crystallinity, it is not suitable for thermoplastic or fiber use. The turning point in polypropylene production was the development of a Ziegler-type catalyst by Natta to produce the stereoregular form (isotactic).

Polypropylene could be produced in a liquid or in a gas-phase process. Until 1980, the vertically stirred bed process of BASF was the only largescale commercial gas phase process.8 In the Union Carbide/Shell gas-phase process (Figure 2), a wide range of polypropylenes are made in a fluidized bed gas phase reactor. Melt index, atactic level, and molecular weight distribution are controlled by selecting the proper catalyst, adjusting operating conditions, and adding molecular weight control agents. This process is a modification of the polyethylene process, but a second reactor is added. Homopolymers and random copolymers are produced in the first reactor, which operates at approximately 70°C and 35 atmospheres. Impact copolymers are produced in the second reactor (impact reactor) after transferring the polypropylene resin from the first reactor. Gaseous propylene and ethylene are fed to the impact reactor to produce the polymers' rubber phase. Operation of the impact reactor is similar to the initial one, but the second operates at lower pressure (approximately 17 atmospheres). The granular product is finally pelletized.

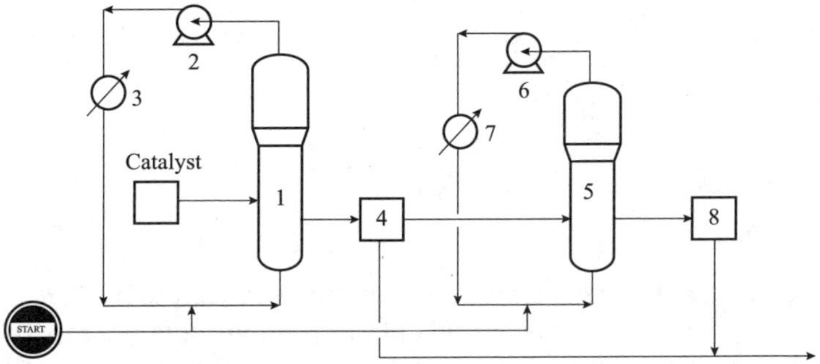

Figure 2 The Union Carbide gas–phase process for producing polypropylene: (1) reactor, (2) centrifugal compressor, (3) heat exchanger, (4) product discharge tank (unreacted gas separated from product), (5) impact reactor, (6) compressor, (7) heat exchanger, (8) discharge tank (copolymer separated from reacted gas).

4 Polyvinyl Chloride

Polyvinyl chloride (PVC) is one of the most widely used thermoplastics. It can be extruded into sheets and film and blow molded into bottles. It is used in many common items such as garden hoses, shower curtains, irrigation pipes, and paint formulations.

PVC can be prepolymerized in bulk to approximately 7%~8% conversion. It is then transferred to an autoclave where the particles are polymerized to a solid powder. Most vinyl chloride, however, is polymerized in suspension reactors made of stainless steel or glass-lined. The peroxide used to initiate the reaction is dispersed in about twice its weight of water containing 0.01%~1% of a stabilizer such as polyvinyl alcohol.

In the European Vinyls Corp. process (Figure 3), vinyl chloride monomer (VCM) is dispersed in water and then charged with the additives to the reactor.19 It is a stirred jacketed type ranging in size between 20~105m^3. The temperature is maintained between 53~70°C to obtain a polymer of a particular molecular weight. The heat of the reaction is controlled by cooling water in the jacket and by additional reflux condensers for large reactors. Conversion could be controlled between 85%~95% as required by the polymer grade.

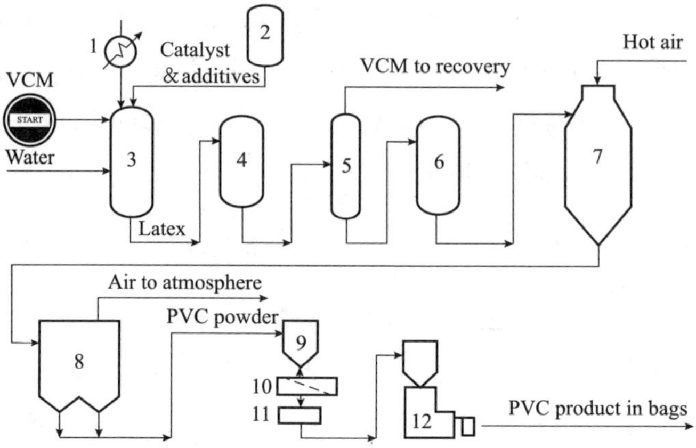

Figure 3 The European Vinyls Corp. process for producing polyvinyl chloride using suspension polymerization: (1) reactor, (2) blow-down vessels (to separate unreacted monomer), (3) stripping column, (4) reacted monomer recovery, (5) slurry centrifuge, (6) slurry drier.

5 Polystyrene

Polystyrene (PS) is the fourth big-volume thermoplastic. Styrene can be polymerized alone or copolymerized with other monomers. It can be polymerized by free radical initiators or using coordination catalysts.

Copolymers of styrene-acrylonitrile (SAN) have higher tensile strength than styrene homopolymers. A copolymer of acrylonitrile, butadiene, and styrene (ABS) is an engineering plastic due to its better mechanical properties (discussed later in this chapter). Polystyrene is produced either by free radical initiators or by use of coordination catalysts. Bulk, suspension, and emulsion techniques are used with free radical initiators, and the polymer is atactic.

SAN (T_g=105°C) is stiffer and has better chemical and heat resistance than the homopolymer. However, it is not as clear as polystyrene, and it is used in articles that do not require optical clarity, such as appliances and houseware materials.

ABS has a specific gravity of 1.03 to 1.06 and a tensile strength in the range of 6 to 7.5×10^3 psi. These polymers are tough plastics with outstanding mechanical properties. A wide variety of ABS modifications are available with heat resistance comparable to or better than polysulfones and polycarbonates (noted later in this section). Another outstanding property of ABS is its ability to be alloyed with other thermoplastics for improved properties. For example, ABS is alloyed with rigid PVC for a product with better flame resistance.

6 Nylon resins

Nylon resins are important engineering thermoplastics. Nylons are produced by a condensation reaction of amino acids, a diacid and a diamine, or by ring opening lactams such as caprolactam. The polymers, however, are more important for producing synthetic fibers.

Important properties of nylons are toughness, abrasion and wear resistance, chemical resistance, and ease of processing. Reinforced nylons have higher tensile and impact strengths and lower expansion coefficients than metals. They are replacing metals in many of their applications. Objects made from nylons vary from extruded films used for pharmaceutical packaging to bearings and bushings, to cable and wire insulation.

Nylon 66 is produced by the reaction of hexamethylenediamine and adipic acid. This produces hexamethylenediammonium adipate salt. The product is a dilute salt solution concentrated to approximately 60% and charged with acetic acid to a reactor where water is continuously removed. The presence of a small amount of acetic acid limits the degree of polymerization to the desired level:

$$n\mathrm{HO\overset{\overset{O}{\|}}{C}(CH_2)_4\overset{\overset{O}{\|}}{C}OH} + n\mathrm{H_2N(CH_2)_6NH_2} \longrightarrow [\mathrm{O\overset{\overset{O}{\|}}{C}(CH_2)_4\overset{\overset{O}{\|}}{C}O}]_n^{2-}\,[\mathrm{H_3N(CH_2)_6NH_3}]_n^{2+}$$

$$\longrightarrow \left[\mathrm{\overset{\overset{O}{\|}}{C}(CH_2)_4\overset{\overset{O}{\|}}{C}NH(CH_2)_6NH}\right]_n + 2n\mathrm{H_2O}$$

The temperature is then increased to 270~300°C and the pressure to approximately 16 atmospheres, which favors the formation of the polymer. The pressure is finally reduced to atmospheric to permit further water removal. After a total of three hours, nylon 66 is extruded under nitrogen pressure.

Caprolactam w-Amino acid

Nylon 6 is produced by the polymerization of caprolactam. The monomer is first mixed with water, which opens the lactam ring and gives *w*-amino acid.

Nylon 4 is produced by ring opening 2-pyrrolidone. Anionic polymerization is used to polymerize the lactam. Cocatalysts are used to increase the yield of the polymer. Carbon dioxide is reported to be an excellent polymerization activator. Nylon 4 has a higher water absorption capacity than other nylons due to its lower hydrophobic property.

7 Thermoplastic Polyesters

Thermoplastic polyesters are among the large-volume engineering thermoplastics produced by condensation polymerization of terephthalic acid with ethylene glycol or 1,4-butanediol. These materials are used to produce film for magnetic tapes due to their abrasion and chemical resistance, low water absorption, and low gas permeability. Polyethylene terephthalate (PET) is also used to make plastic bottles (approximately 25% of plastic bottles are made from PET). Similar to nylons, the most important use of PET is for producing synthetic fibers.

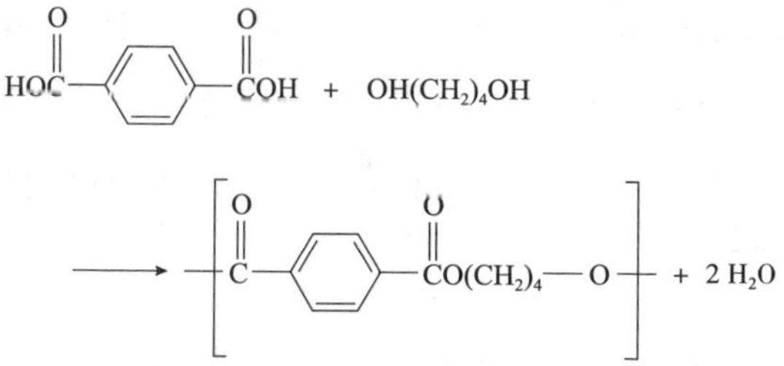

Polybutylene terephthalate (PBT) is another thermoplastic polyester produced by the condensation reaction of terephthalic acid and 1,4-butanediol.

8 Polycarbonates

Polycarbonates (PC) are another group of condensation thermoplastics used mainly for special engineering purposes. These polymers are considered polyesters of carbonic acid. They are produced by the condensation of the sodium salt of bisphenol A with phosgene in the presence of an organic solvent. Sodium chloride is precipitated, and the solvent is removed by distillation:

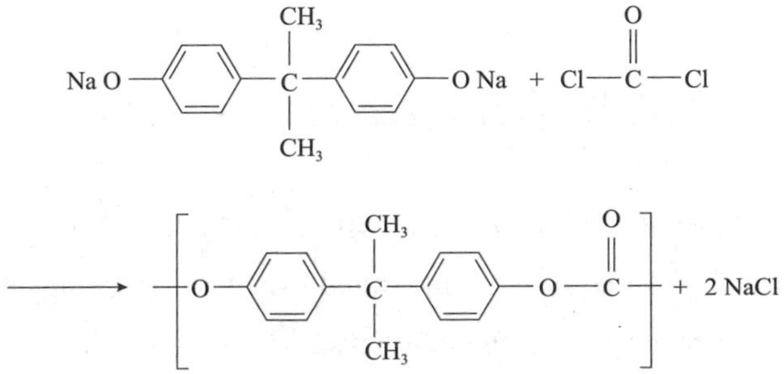

Another method for producing polycarbonates is by an exchange reaction between bisphenol A or a similar bisphenol with diphenyl carbonate:

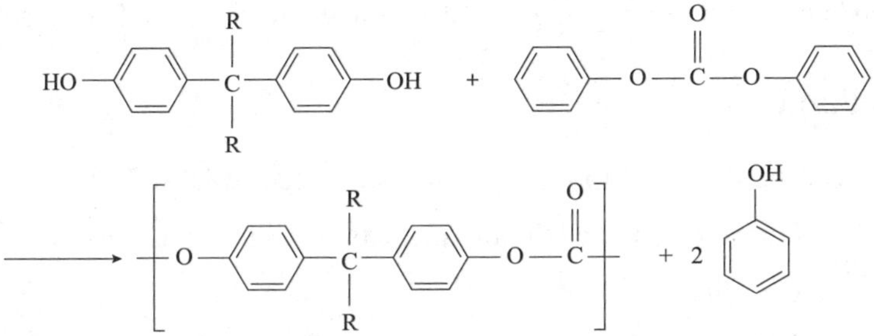

Diphenol carbonate is produced by the reaction of phosgene and phenol. A new approach to diphenol carbonate and non-phosgene route is by the reaction of CO and methyl nitrite using Pd/alumina. Dimethyl carbonate is formed which is further reacted with phenol in presence of tetraphenox titanium catalyst. Decarbonylation in the liquid phase yields diphenyl carbonate.

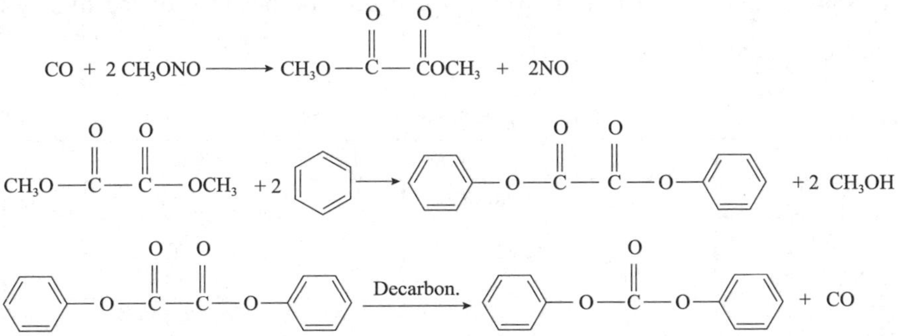

Selected from "*Chemistry of Petrochemical Processes* (2nd edition)", Sami Matar, Gulf Publishing Company, 2000.

Olefins and Aromatic Plants

The closest refiners come to becoming petrochemical companies happens when they build ethylene plants. That's the reason why integrated oil/chemical companies built so many large ethylene plants: they bridged the gap between the two. Ethylene plants are better called olefins plants, but are variously referred to as ethylene crackers (a misnomer—they don't crack ethylene), steam crackers (another misnomer, although steam is mixed with the feed) or something-crackers, with the prefix denoting the feed, as in ethane cracker.

1 Olefin Plants

Olefin plants can be designed to crack a number of feedstocks. They usually fall into the following categories: Ethane, Ethane/propane mix, Propane, Butane, Naphtha or natural gasoline, Gas oil.

The original olefin plants were built to produce ethylene to supply the growing appetite of the petrochemical industry for this basic building block. The propylene coming from an olefin plant attracted less interest. Propylene in abundant amounts was always readily available to be filched from refinery alkylation plants. Many early U.S. olefins plants were designed to crack ethane or ethane and propane, first because ethane was abundantly available from natural gas and second, because of the high yield of ethylene from ethane (Table 1).

Table 1. Ethylene plant yields

	Pounds of Product per Pound of Feed				
	Ethane	Propane	Butane	Naphtha	Gas oil
Ethane/ethylene	0.77	0.40	0.36	0.23	0.18
Propane/propylene	0.01	0.18	0.20	0.13	0.14
Butanes/butylenes	0.01	0.02	0.05	0.15	0.06
Butadiene	0.01	0.01	0.03	0.04	0.04
Fuel gas	0.20	0.38	0.31	0.26	0.18
Gasoline	—	0.01	0.05	0.18	0.18
Gas oil	—	—	—	0.01	0.12
Fuel oil	—	—	—	—	0.10

In other parts of the world where natural gas resources were scarce (Europe and Asia), naphtha emerged as the preferred flavor of feed. The demand for gasoline as a percent of crude was always lower outside the U.S. That left lots of naphtha available to the petrochemical industry at relatively attractive prices. A few olefin plants have been built to crack gas oil, producing as by-products some high-octane gasoline blending components. Those olefin plants, as large as medium-sized refineries, were integrated into refineries and produce a significant amount of the gasoline blending components and other feedstocks.

(1) The Process

Ethane/propane crackers, as shown in the schematic, Figure 1, are the simplest design, but they demonstrate the fundamentals. The ethane and propane can be fed separately or as a mixture to the cracking furnaces where the short residence time (a tenth of a second) and 1200~1600°F temperature yield a high volume of ethylene. The quench downstream of the cracking furnaces drops the temperature quickly and stops the cracking process from taking the feeds all the way to methane.

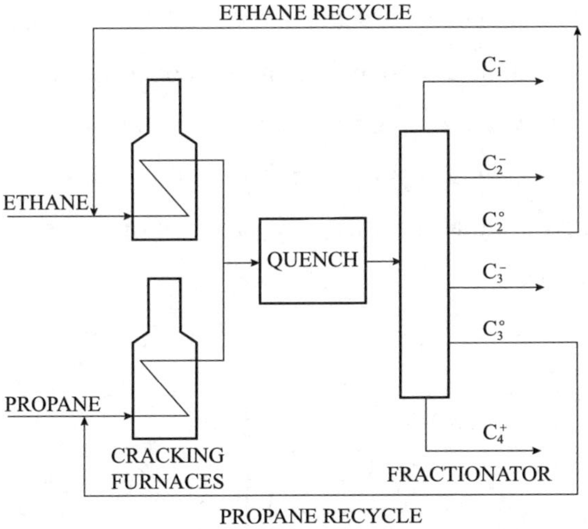

Figure 1 Schematic ethane–propane cracker

In one pass through the cracking furnace, only about sixty to seventy percent of the ethane and propane disappear (crack). Downstream in the product fractionator, the uncracked ethane and propane are split out and recycled to the feed. Generally, the ethane is recycled to extinction, but some propane leaves with the propylene.

Figure 2 shows a better, cartoon view of an ethane cracker, but still leaves out most of the auxiliary equipment. Other than the operating conditions and feeds, plus a few bells and whistles, olefin plants are just plain old thermal crackers.

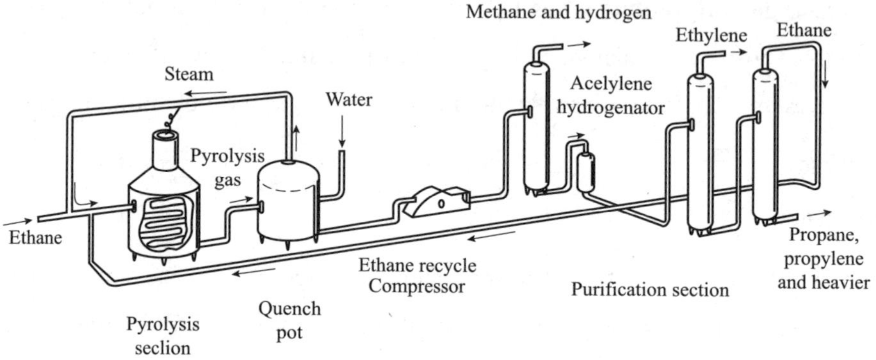

Figure 2 Ethane cracker

The olefin plants designed to crack heavier feedstocks need extra separation facilities. Table 1 shows that ethylene and propylene make up only about thirty-five percent of the yield from cracking naphtha and gas oil. The rest has to be separated by a series of columns and maybe a cryogenic unit to recover the ethylene. Figure 3 greatly simplified the scheme by showing only a single fractionator, but Figure 4 shows the elaborate separation scheme necessary to sort out all the steams. The yield in Table 1 from all the feeds assumes that the ethane coming out of the reactors is recycled to an ethane furnace, as shown.

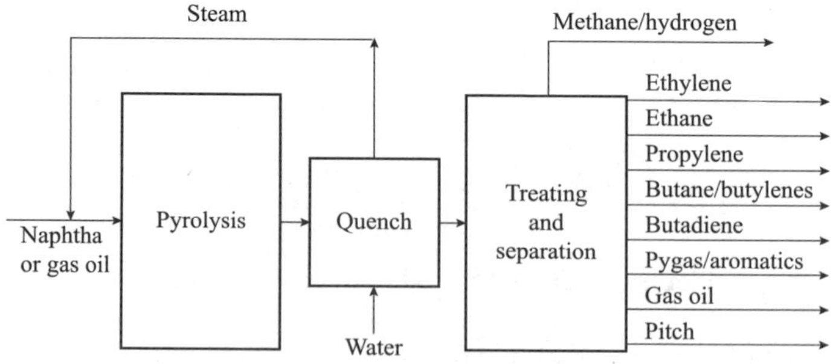

Figure 3 Schematic naphtha-gas oil cracker

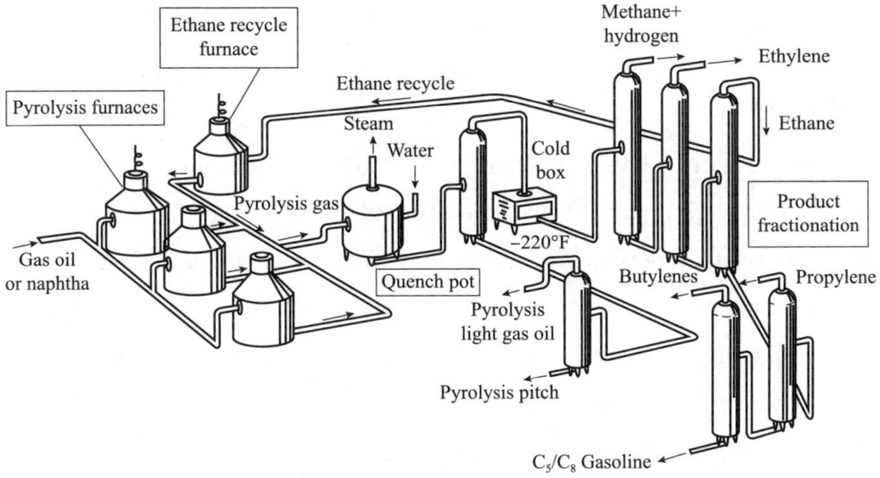

Figure 4 Naphtha–gas oil cracker

(2) Products

Ethylene is normally sold as a ninety-nine+ percent purity stream. Deliveries are almost always by pipeline. Propylene is bought and sold in three different commercial grades that differ by the accompanying amount of propane mixed in. Polymer grade propylene is ninety-seven to ninety-nine+ percent propylene; chemical grade is ninety-two percent to ninety-five percent; refinery grade is fifty to sixty-five percent. Refinery grade propylene is the propane/propylene mixture coming from cat crackers, the biggest source of propylene in a refinery. Chemical grade propylene comes out of naphtha and gas oil cracking in an olefins plant at about ninety-two percent with eight percent propane. Either of these two grades can be fractionated to get the polymer grade.

A commercial petrochemical product, butadiene, turns up in the olefin plant outturns but is rarely seen in the refinery sector. This chemical is in the di-olefin family and has two double bonds with the formula C_4H_6. The two double bonds make it doubly reactive, especially for creating plastics and rubber compounds (elastomers). This extra reactivity, however, makes it prone to gum formation and other undesirable reactions when it is used in gasoline or as feed to an alky plant. For those applications, a C_4 stream with the butadiene can be hydrotreated over a catalyst, turning the butadiene into butylene or butane.

2 Aromatics

(1) Applications

Removal of aromatic compounds can be desirable for different reasons. Either the aromatics have detrimental effects on the quality of the mixture they're in, or the aromatics are worth more if they're separated than if they're not. You have already seen a couple of examples:

(a)Gasoline now has a max benzene spec. (b) Aromatic compounds in kerosene can cause unacceptable smoke points.

Other more specialized applications include:

(a) Kerosene range solvents that are aromatics-free or aromatics-laden and have various industrial applications; (b) Separated benzene, xylene, and toluene, which have numerous chemical applications; (c) Removing aromatics from heavy gas oil stocks can improve the lubricating oil characteristics.

(2) Processes

The solvent recovery process is based on the ability of certain compounds to dissolve selectively certain classes of other compounds. In this case, certain solvents will dissolve aromatics but not paraffins, olefins, or naphthenes. The reasons the process works are another story, i.e., a different book that I haven't attempted. (It can't be nontechnical.)

The first co-requisite that makes this approach successful is that the solvent with the extracted compounds dissolved in it readily separates itself from the starting hydrocarbon mixture. The second co-requisite is that the solvent and the dissolved, targeted extract can easily be split in a fractionator. Take kerosene as an example, one that has a lot of aromatic compounds in it. To half a beaker of kerosene, add a half beaker of a solvent—in this case, liquid sulfur dioxide. After mixing, the liquid will separate into two phases with the kerosene on the bottom and the sulfur dioxide on top. The kerosene on the bottom will fill less than half the beaker. The sulfur dioxide, because the aromatic compounds have dissolved in it, will take up more than half.

If the sulfur dioxide is poured off, the aromatic compounds can be "sprung" from it by simple distillation. This two-step process is batch solvent processing.

Knowing how a simple batch process works, a continuous flow process is easy to conceptualize. Figure 5 shows a three column system. The feed is introduced as a vapor into the lower part of a column with a labyrinth of mixers inside. (Sometimes the mixers are mechanically moved, such as in a rotating disc contactor.) The solvent is introduced as a liquid near the top. The solvent works its way towards the bottom of the vessel, dissolving the extract as it goes by. The rest of the hydrocarbon, which emerges from the top, is called raffinate.

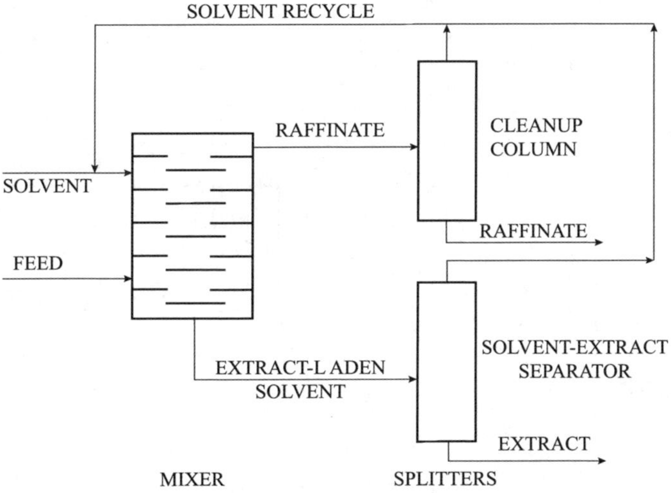

Figure 5 Solvent extraction

Two columns handle the streams coming out of the mixer. One column cleans up any small amounts of solvent that may have followed along with the raffinate. The other column separates the solvent and the extract. The solvent from both columns is recycled to the top of the mixer.

Some of the solvents used in various applications are listed below.

(a) Kerosene treating: liquid SO_2, Furfural. (b) Lubricating oil treating: liquid SO_2, Furfural, phenol, propane (separates paraffins from asphaltenes). (c) Gasoline: Sulfolane, phenol, acetonitrile, liquid SO_2.

A good solvent has a boiling temperature sufficiently higher than the targeted extract so that separating the two in a fractionating column is easily accomplished.

(3) Benzene and Aromatics Recovery

The most widespread application of solvent extraction, especially in refineries, is for BTX recovery, especially for benzene. To make the process efficient, the feed to the process is pared down to an aromatic concentrate by making a heart cut from a reformate or straight run gasoline stream as shown in Figure 6. The aromatics concentrate then has a higher benzene percent, making the extraction process more efficient.

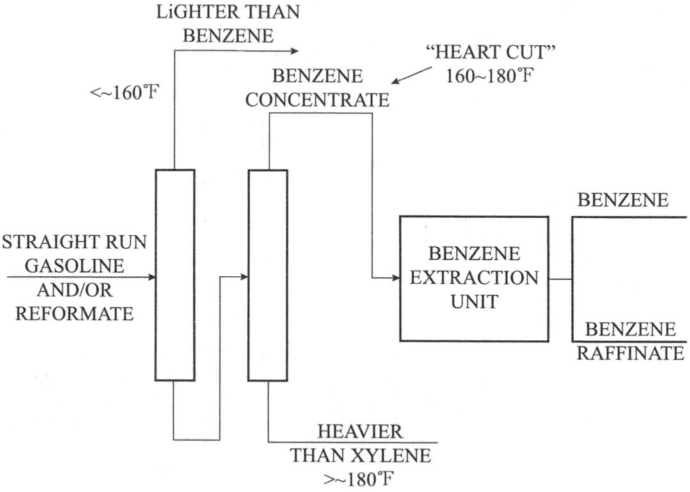

Figure 6 Solvent Recovery of Aromatics

One bit of nomenclature is often misleading. Benzene raffinate contains no benzene. It is the leftover of the aromatics concentrate after the target (benzene) is removed. As a result, benzene raffinate becomes an acceptable gasoline blending component, and the separated benzene becomes a chemical feedstock.

3 Refinery Interactions

Streams go back and forth when a refinery integrates with an olefins plant(Figure 7) . Olefin plants sometimes provide a home for a number of "junk "streams in a refinery. For example, the dry gas stream from a cat cracker is usually sent to the fuel system, even though it contains ethane, ethylene, some propane, and propylene. These components can be recovered by commingling them with the furnace outturns of the olefin plant as they head for the separation section.

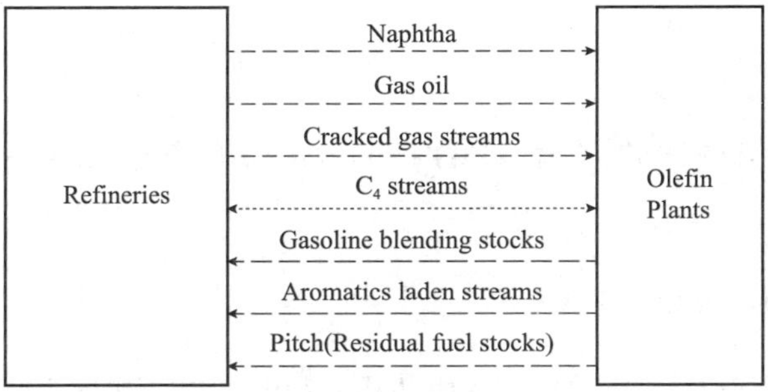

Figure 7 Refinery–Olefin Plant Exchanges

As another example, some of the naphtha in the gasoline pool are of such low octane they get shut out of the gasoline pool as a blending component. A normal pentane (RON of sixty-two) or normal hexane (RON of twenty-six) stream in refineries that have no isomerization unit could be examples. These streams can make attractive feedstocks for refiner-owned olefins plants, turning trash into treasures.

Sometimes the price of normal butane is low enough that it is an attractive olefin plant feedstock. On the outturn side of the olefin plant, refiners find some other complementary opportunities, too. Besides the gasoline range products, the butanes and butylenes can generally find a home in the refinery processes.

Benzene, toluene, and xylenes can go in both directions. An olefins plant cracking naphtha or gas oil will have a gasoline stream rich in aromatics. (The benzene ring is very resistant to cracking). So an aromatics recovery unit could be placed in the refining or petrochemical part of a complex.

Selected from Chapter 19 of "*Petroleum Refining in Nontechnical Language* (5th edition)", William L Leffler, Matthew Dresher, 2020.

Catalysts for Petrochemical Processing

1 Hydrocracking catalysts

Hydrocracking catalysts have a cracking function and a hydrogenation and dehydrogenation function (Figure 1). The cracking function is provided by an acidic support, whereas the hydrogenation-dehydrogenation function is provided by active metals. The acidic support can be (a) amorphous oxides (e.g. silica-alumina), (b) a crystalline zeolite (mostly modified Y zeolite) plus binder (e.g. alumina) or (c) a mixture of crystalline zeolite and amorphous oxides. Cracking and isomerization reactions take place on the acidic support.

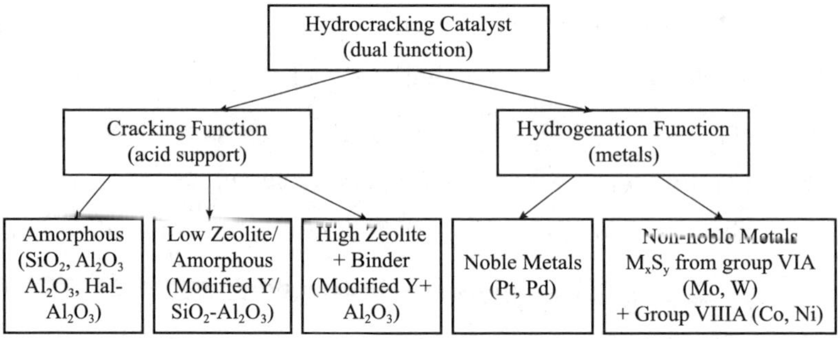

Figure 1 Classification of hydrocracking catalyst

The metals providing the hydrogenation-dehydrogenation function can be noble metals (palladium, platinum) or non-noble metal sulphides from Group VI.A (molybdenum, tungsten) and group VIII.A (cobalt, nickel). These metals catalyze the hydrogenation of the feedstock, making it more reactive for cracking and heteroatom removal and reducing the coking rate. They also initiate the cracking by forming a reactive olefin intermediate via dehydrogenation. The ratio between the catalyst's cracking function and hydrogenation function can be adjusted to optimize activity and selectivity. The relative strength of different hydrogenation components and cracking (acid) components in hydrocracking catalysts are shown in Table 1.

Table 1. Bifunctional catalyst strength for hydrogenation and cracking

Hydrogenation function	Co/Mo	Ni/Mo	Ni/W	Pt (Pd)
	x	xx	xxx	xxxx
Cracking function	Al_2O_3	Al_2O_3–hallogen	SiO_2–Al_2O_3	zeolite
	x	x	xxx	xxxx

x represents order of strength.

Catalysts with amorphous supports are still in commercial use, primarily where maximizing the production of middle distillates or conversion to lube oil blending stock are the objective. Amorphous hydrocracking catalysts contain primarily amorphous silica-alumina. The hydrocracking catalysts described so far are used primarily for hydrocracking gas oils and FCC cycle oils. For hydroprocessing residues, amorphous hydrocracking catalysts as well as specially designed hydrotreating catalysts and iron-containing catalysts are used.

Catalysts used fo rmild hydrocracking have a composition similar to that of hydrotreating catalysts. They consist of Group VI and VIII non-noble metals supported on g-alumina. The metals used are cobalt, nickel, molybdenum, and tungsten in sulphided form. Under mild process conditions, gas oil hydrocracking catalysts may also be used for mild hydrocracking. Dewaxing catalysts usually consist of a hydrogenation metal (Pt, Pd, Ni) supported on a medium-pore zeolite (e.g. ZSM-5) combined with a binder, commonly alumina. A zeolite catalyst Y type is shown in Figure 2. The reacting molecules pass inside the cages and are brought in close contact for reaction. The zeolite channels (or pores) are microscopically small and in fact have molecular size dimension such that they are often termed "molecular sieves".

The composition of a hydrocracking catalyst depends on the final product requirements and mode of the operation (one or two-stage). A guideline of catalysts composition is shown in Table 2. In the first stage of a two-stage operation, a hydrocracking hydrogenation catalyst is usually used for hydrodesulphurization, hydrodenitrogenation and aromatics removal. In the second stage a hydrocracking catalyst is used for a hydrocracking and hydroisomerization. In Table 3, the types of catalysts used for mild and high conversion hydrocracking are shown along with FCC and hydrotreating catalysts. Residue conversion processes (fixed, moving and ebulated beds) use supported palletized catalysts of bifunctional composition. Slurry reactors employ dispersed catalysts.

Table 2. Types of catalyst used in different hydrocracking processes

Application type	Stages	Hydrotreating function			Acid function	
		Pd	Ni-Mo	Ni-W	Y zeolite	SiO_2-Al_2O_3 amorphous
Naphtha	One	xxx	x	x	xxx	—
Naphtha	Two	xxx	—	—	xxx	—
Kerosene	One	—	xxx	x	xxx	—
Kerosene	Two	x	x	x	x	x
Diesel oil	One	—	x	x	x	x
Diesel oil	Two	—	x	x	—	x
Lubricant oil	One	—	x	x	x	x

Table 3. Catalysts used for VGO (vacuum gas oil) treatment

High conversion hydrocracking	Bifunctional	A: Amorphous Silica-Alumina, zeolite
		H: Pd, Pt, Ni (Co), Mo (W)
Mild hydrocracking	Mono-functional or bifunctional	A: Amorphous Silica-Alumina, Alumina
		H: Ni (Co), Mo (W)
FCC	Mono-functional	A: Amorphous Silica-Alumina+zeolite
Hydrotreating	Mono-functional	H: Ni (Co), Mo (W)

2 Reforming catalysts

The catalyst used for reforming is a bifunctional catalyst composed of platinum metal on chlorinated alumina. Platinum acts as the center for the dehydrogenation reaction, and chlorinated alumina acts as an acidic site to promote structure changes, such as cyclization of paraffins and isomerization of the naphthenes.

Recently additional elements have been added to platinum to promote additional properties for the catalyst. Iridium (Ir) is added to boost activity, Rhenium (Re) is added to operate at lower pressures and Tin (Sn) is added to improve yield at low pressures. The use of Pt/Re is now most common in semi-regenerative (SR) processes with Pt/Sn is used in moving bed reactors. The quantity of chlorine used is approximately 1wt% of the catalyst and the quantity of platinum is from 0.2wt% to 0.6wt%. Impurities that might cause deactivation or poisoning of the catalyst include: coke, sulphur, nitrogen, metals and water. Because of these problems, the reformer feed has to be severely hydrotreated to remove most of these impurities, and the reformer should be operated at high temperature and low pressure to minimize coke deposition.

Paraffin and naphthene dehydrogenation reactions are very rapid and usually occur in the first reactor. The isomerization of paraffin and naphthenes is fast, whereas hydrocracking is slow and takes place in the last reactor.

Bimetallic catalysts such as Pt/Re were found to have better stability, increased catalyst activity, and selectivity. Trimetallic catalysts of noble metal alloys are also used for the same purpose. The increased stability of these catalysts allowed operation at lower pressures. A ratio of 0.5 or less for Pt/Re in the new generation catalysts versus 1.0 for the older ones can tolerate much higher coke levels. Reforming units can perform similarly with higher coke levels (20%~25% versus 15%~20%). These catalysts can tolerate higher sulfur naphtha feeds (>1ppm). Higher profitability may be realized by increasing the cycle length.

3 Cracking catalysts

The main catalyst which is used in a FCC reactor is the zeolite type. It is in a powderform with an average particle size of 75mm and an average surface area of 800m^2/g. It has a crystalline structure of aluminosilicates. A matrix is added to the zeolite which acts as a binder and filler.

The main active component in the catalyst is the *Y*-Zeolite. It is a crystalline structure of aluminosilicates which has the *Y*-faujasite structure (Figure 2). The highest pore size in the Y-faujasite structure is 8Å, which is called the super cage. It can allow some C_{18}~C_{25} mono-, di- and tri-nuclear aromatics present in the VGO to pass.

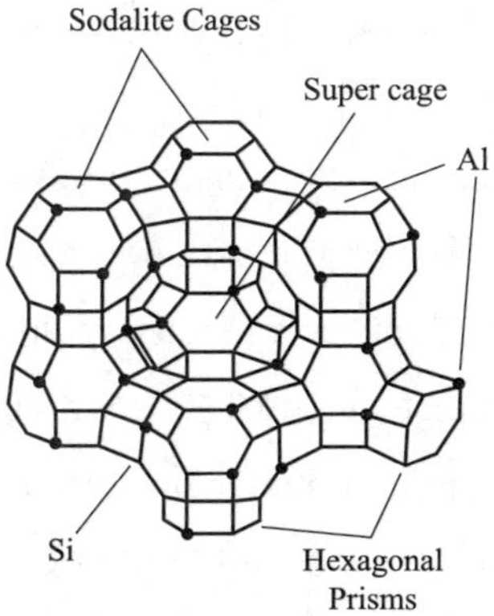

Figure 2 Y-type zeolite catalyst

The four oxygen anions in the tetrahedron are balanced by the +4 oxidation state of the silicon cation, while the four oxygen anions connecting the aluminum cation are not balanced. This results in –1 net charge, which should be balanced. Metal cations such as Na^+, Mg^{2+}, or protons (H^+) balance the charge of the alumina tetrahedra.

Bronsted acid sites in HY-zeolites mainly originate from protons that neutralize the alumina tetrahedra. When HY-zeolite (*X*- and *Y*-zeolites are cracking catalysts) is heated to temperatures in the range of 400~500°C, Lewis acid sites are formed.

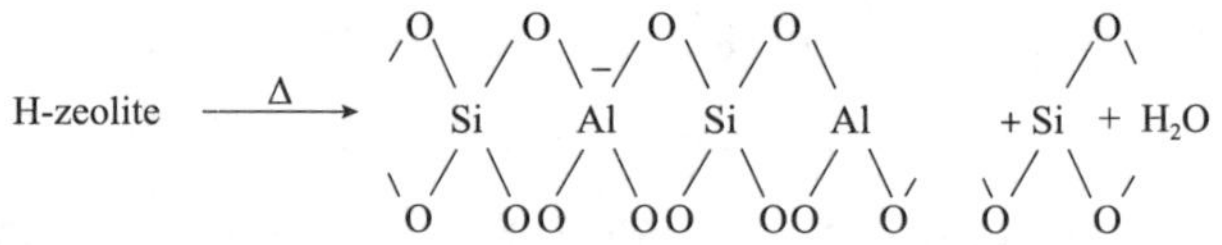

A Lewis acid site

Zeolites as cracking catalysts are characterized by higher activity and better selectivity toward middle distillates than amorphous silica-alumina catalysts. This is attributed to a greater acid sites density and a higher adsorption power for the reactants on the catalyst surface.

The higher selectivity of zeolites is attributed to its smaller pores, which allow diffusion of only smaller molecules through their pores, and to the higher rate of hydrogen transfer reactions. However, the silica-alumina matrix has the ability to crack larger molecules. In a set of experiments using gas oil and rare earth zeolite/silica-alumina, the yield of gasoline increased when the matrix was used before the zeolite. This was explained by the mechanism of initial matrix cracking of large feedstock molecules to smaller ones and subsequent zeolite cracking of the smaller molecules to products.

Aluminum distribution in zeolites is also important to the catalytic activity. An inbalance in charge between the silicon atoms in the zeolite framework creates active sites, which determine the predominant reactivity and selectivity of FCC catalyst. Selectivity and octane performance are correlated with unit cell size, which in turn can be correlated with the number of aluminum atoms in the zeolite framework.

Deactivation of zeolite catalysts occurs due to coke formation and to poisoning by heavy metals. In general, there are two types of catalyst deactivation that occur in a FCC system, reversible and irreversible. Reversible deactivation occurs due to coke deposition. This is reversed by burning coke in the regenerator. Irreversible deactivation results as a combination of four separate but interrelated mechanisms: zeolite dealumination, zeolite decomposition, matrix surface collapse, and contaminationby metals such as vanadium and sodium.

Pretreating the feedstocks with hydrogen is not always effective in reducing heavy metals, and it is expensive. Other means that proved successful are modifying the composition and the microporous structure of the catalyst or adding metals like Sb, Bi or Sn, or Sb-Sn combination. Antimony organics have been shown to reduce by 50% gas formation due to metal contaminants, especially nickel.

In the cracking of long chain paraffins, another type of high silica zeolite is added. This zeolite is called ZSM-5 and is used to improve octane number and it is composed of zig-zag channel systems. Figure 3 shows how hexane will easily enter the catalyst pore. Then hexane cracks into propane, while iso-hexane and benzene do not enter the pores. Thus the unreacted stream is enriched with iso-paraffins and aromatics, which contribute to an increase in the octane number.

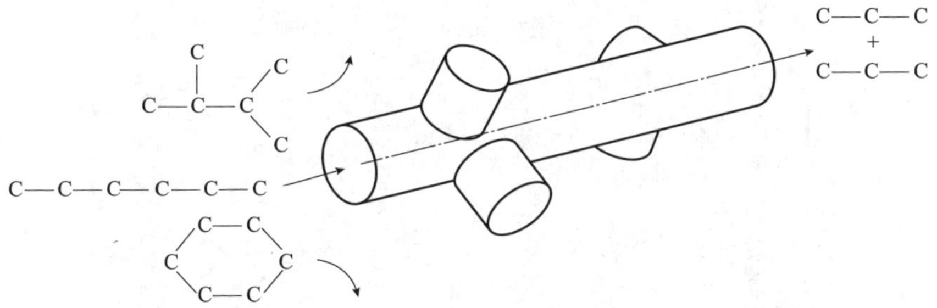

Figure 3 Schematic representation of shape–selective cracking with ZSM–5 zeolite

The matrix is added to the zeolite to increase the body of the catalyst and add some improved properties. Three types of substances constitute the matrix:

(a) A binder is added as a glue which is then added to the catalyst to provide cohesion for zeolite particles.

(b) A filler is added to make up the body of the catalyst. It is usually a clay (Kaoline). The function of the filler and binder is to provide physical integrity (density and attrition resistance).

(c) Additives, such as a small amount (ppm) of metal, are added to the catalyst to promote the combustion of CO to CO_2 in the regenerator. Metallic oxides are added to fix SO_x on the catalyst. The sulphur is recovered as H_2S in the reactor. An addition of 5% ZSM-5 zeolite will lead to an increase of one research octane number (RON).

Selected from “*Chemistry of Petrochemical Processes* (2nd edition)”, Sami Matar, Gulf Publishing Company, 2000.

Selected from “*Fundamentals of Petroleum Refining*”, Taher A. Alsahhaf, Amal Sayed Elkilani, M.A. Fahim, Elsevier, 2010.

Chapter 6

Batteries
（电池）

Part A Comprehensive Reading

Battery Technologies

1 Background

Batteries are devices that transform chemical energy into electricity. Every battery has two terminals or electrodes, i.e., the positive cathode and the negative anode. As the device is switched on, the chemical reactions are started and the electrons produced travel from the anode to the cathode and electrical work is generated. This is the discharge process. During charging the operation is reversed.

As a single unit cell is limited in terms of voltage and **capacity**, a battery commonly consists of several cells that are electrically connected. By connecting a number of cells in series, a battery pack is formed. The voltage is then the sum of the voltages of each cell while the current is the same. The electrochemical cells may also be connected in **parallel** and then the voltage is the same for all cells while the current is the sum of each current.

A primary battery is disposable because its electrochemical reactions cannot be reversed. A primary battery is sometimes called a single-use battery. On the other hand, a secondary battery is rechargeable because the electrochemical reaction can be reversed by applying a certain voltage to the battery in the opposite direction of the discharge.

The battery capacity Q is the product of the current and time, i.e., $Q = I \times t$ and the unit is ampere hour (Ah) or milliampere hour (mAh). The product of the battery voltage and the rated ampere hours is called the Watt hours of the battery and is calculated as Watt Hours (W·h)=Battery Voltage×Ah ($V \times I \times t$). The specific energy density is defined as the amount of energy stored per unit mass, i.e., W·h/kg. The **energy density** is defined as the amount of energy stored per unit volume, i.e., W·h/m^3. The property power density is sometimes used and this is defined as the maximum rate of energy discharge per unit mass or volume. The theoretical capacity of an electrode is related to the charge transferred in a specific reaction. It is derived from Faraday's law.

$$Q_{\text{th}} = n_{\text{e}} \cdot \frac{F}{M_{\text{W}}}$$

where Q_{th} is the theoretical specific capacity in Ah/kg, n_{e} is the number of moles of transferred electrons, F the Faraday constant and M_{w} is the molecular weight of the electrochemical active material (kg/kmol).

The components of a battery might be described as active ones and these are the electrodes. The non-actives ones are then the electrolyte, separator, current collectors and the casing.

The electrodes are manufactured in electrically and ionically conducting materials. Metal electrodes permit the electrochemical reactions to occur only at the outermost layers and such are called blocking electrodes. For so-called insertion or non-blocking electrodes, the redox reactions occur at the surface as well as in the bulk of the electrodes and such are used in rechargeable cells.

The electrolyte is commonly a solution of one or several salts dissolved in one or several solvents. The main task of the electrolyte is to conduct ions. Also it should prohibit short circuits.

The separators improve the mechanical strength of the electrolyte and help to avoid internal short circuits. Commonly the separators are in a porous material with high ion conductivity and good electronic insulation.

The current collectors are used to provide efficient transfer of electrons. Usually thin foils of copper or aluminum are uses. The current collectors should also be able to remove heat from the electrodes.

The casing aims to provide mechanical stability and prevents influence from the outside. Also id protects a liquid electrolyte from being evaporated. The casing material is commonly of plastic or metallic material.

C-rate is a measure of the discharge or charge rate relative to the maximum cell capacity. 1C rate is a discharge current discharging the cell within 1h.

The single-use or primary batteries are of alkaline or lithium type and have very low self-discharge rate for a long shelf life. They are widely available and are cheaper than rechargeable batteries. However, they must be disposed after being fully discharged. The alkaline batteries are employing the chemical reaction of zinc (anode) and magnesium oxide (cathode) and use an alkaline electrolyte of potassium hydroxide to generate electric current. The advantage of alkaline batteries over other primary batteries is their higher energy density.

The rechargeable batteries are built to be recharged many times, up to or more than 500 times. Commonly they are of nickel-metal hydride or lithium-ion type. However, there are manufacturers providing specially designed battery cells to make alkaline batteries rechargeable as well. The rechargeable batteries generate less waste than primary batteries and offer better long-term value but they are more expensive than the primary ones.

Nickel-metal hydride batteries may also come as pre-charged rechargeable batteries. These can then be inserted and used directly in devices. They also show lower self-discharge rate than the standard nickel-metal hydride batteries.

2 Lead–Acid Batteries

As lead batteries are mentioned, it is common to think about a car battery which is used as a starting battery. It delivers a short burst of high power to the starter and ignition system to start the combustion engine. It also supplies extra power when the electrical load of the vehicle exceeds that supplied from the charging system. Starter batteries have a low internal resistance that is achieved by having additional plates for maximum surface area. The plates are thin and the lead is supplied in a sponge-like form which expands the surface area further. The plate thickness is less important as the discharge is short and the battery is recharged while driving. Figure 1 shows a principle sketch of a lead-acid battery. The components are clearly indicated.

Lead batteries are also used as deep-cycle batteries on, e.g., boats or campers, where they are used to the accessories like trolling motors, winches or lights. Then the batteries deliver a lower but steady level of power for a much longer time period than for a starting battery. Such batteries are also used as standby batteries to supply power in events of power outage. Another usage is as motive batteries to power the electric motor of, e.g., forklift trucks. The deep-cycle batteries have thick plates for improved cyclic abilities.

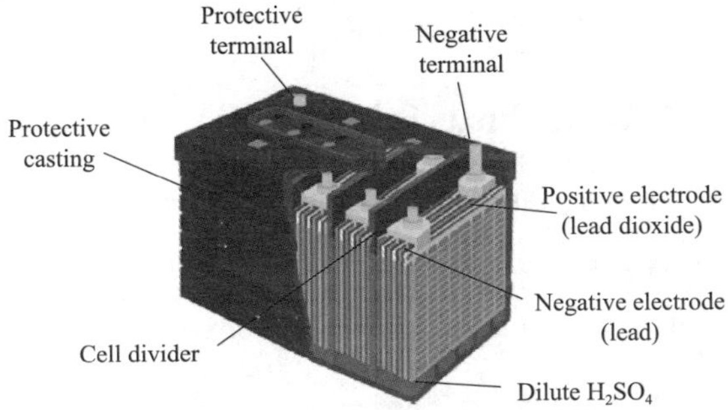

Figure 1 Sketch of a lead–acid battery

Lead batteries may have dual purposes and are then designed to serve a balanced combination of starting and deep cycle service. Depending on the depth of discharge (DoD) the number of cycles varies. The deep-cycle ones allow many more cycles than the starter batteries.

The electrodes are lead and lead-oxide. The electrolyte is a mixture of sulfuric acid (H_2SO_4) and water. The battery is rechargeable.

During the discharge process lead (Pb) is the anode or negative electrode and lead oxide (PbO_2) constitutes the cathode or positive electrode.

The reaction at the anode is:

$$Pb + H_2SO_4^- \longrightarrow PbSO_4 + H^+ + 2e^-$$

The reaction at the cathode is:

$$PbO_2 + 3H^+ + HSO_4^- + 2e^- \longrightarrow PbSO_4 + 2H_2O$$

The overall reaction is:

$$PbO_2 + Pb + H_2SO_4 \longrightarrow 2PbSO_4 + 2H_2O$$

The cell potential is calculated as:

$$E_{cell}=E_+-E_-$$

It should be noted that E_+ and E_- are the standard reduction potentials of the reactions at the half cells. Hence the reduction potential given for the reaction at the anode (Pb) is actually negative. One can find $E_-=-0.356$V. Similarly, for the cathode one finds $E_+=1.685$ V. The cell voltage then becomes:

The disadvantages of the Pb-acid technology have been stated as the heavy weight, the usage at low state-of charge levels (SOCs) and insufficient charge acceptance. In addition lead is toxic. The specific energy is low but the specific power is high. Also the lead-acid batteries provide low self-discharge.

3 Nickel–Metal Hydride Batteries

Nickel Metal hydride, NiMH batteries and cells have a widespread use nowadays and are rechargeable. The NiMH batteries offer very similar characteristics as the NiCd batteries but without the disadvantages of adverse environmental effects that the NiCd cells are associated with.

The NiMH cells have the following main elements: (1) The positive electrode of the NiMH battery is nickel hydroxide. This electrode only exchanges a proton in the charge-discharge reaction. (2) The alkaline electrolyte in the NiMH cell is an aqueous solution of about 30% potassium hydroxide (KOH) which has very high ion conductivity. The electrolyte commonly keeps a constant concentration during the charge-discharge cycle. This is of importance as the electrolyte concentration is a main contributor to the cell resistance. (3) Hydrogen is the active material in the negative electrode but hydrogen is stored as a metal hydride which constitutes the negative electrode. The amount of hydrogen in the metal hydride is in the range of $1e^2$% based on weight. Figure 2 shows a principle sketch of a NiMH cell. The reactions taking place during discharge are as (charging reversed from right to left).

Negative electrode: $MH+OH^- \longrightarrow M+H_2O+e^-$

Positive electrode: $NiOOH+H_2O+e^- \longrightarrow Ni(OH)_2+OH^-$

Overall net reaction: $NiOOH+MH \longrightarrow Ni(OH)_2+M$

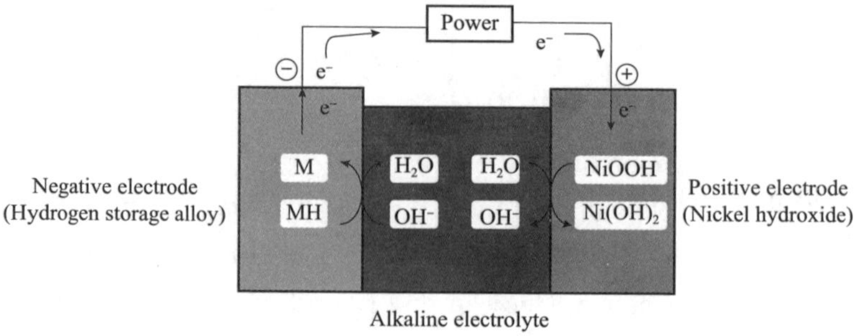

Figure 2 Principle operation of a NiMH cell

NiMH batteries are rated at 1.2~1.3V. These batteries have replaced NiCd batteries and commonly available in AA size. They are often used in digital cameras, in some electric vehicle batteries and in high-drain devices mainly due to their low internal resistance. The energy density is approaching that of lithium-ion batteries. Low self-discharge cells have been developed but these have lower capacity than the standard NiMH cells because the needed thicker separators have larger volume.

4 Lithium Batteries

The most common lithium batteries are lithium metal, lithium-ion and lithium-ion polymer. More recently lithium-oxygen and lithium-sulfur batteries have entered the market. Lithium-ion batteries are today the most frequently used rechargeable batteries and appear in laptops, cell phones, digital cameras and hybrid and electric vehicles. In general, they possess a low weight, high energy density and power density. The cell potential is usually in the range of 2.5~4.5 V.

(1) Lithium Metal Batteries

These batteries have an operating temperature between 80℃ and 120℃ and might be attractive for electric vehicles. Metallic lithium is used as the negative electrode while a lithium insertion material is used as the positive electrode. The electrolyte is polymer based. The presence of metallic lithium at the negative electrode enables the cell to be assembled in the charged or discharged state. Most lithium-metal batteries are non-rechargeable and find applications in film cameras.

(2) Lithium-Ion and Lithium-Ion Polymer Batteries

The concept of lithium-ion batteries was developed to overcome the safety problems associated with cells of lithium metal. However, the energy density is less than for lithium metal batteries. The negative electrode is here based on carbon in the form of graphite or hard carbon. Then, the positive electrode must contain lithium and commonly an insertion material which can allow **insertion** of lithium in one, two or three dimensions is used. Common positive electrode materials are lithium containing transition metal oxides where the metal is Co, Ni, Fe or Mn. The electrolyte in lithium-ion batteries is a lithium salt solvated in an organic solvent. A lithium-ion battery cell has a normal single cell voltage of 3.6V, which is fixed by the battery chemistry. The self-discharge is low and less than for nickel-based batteries. However, the manufacturing is expensive and higher than for, e.g., nickel-cadmium batteries. Figure 3 depicts the principle operation of Li-ion cell.

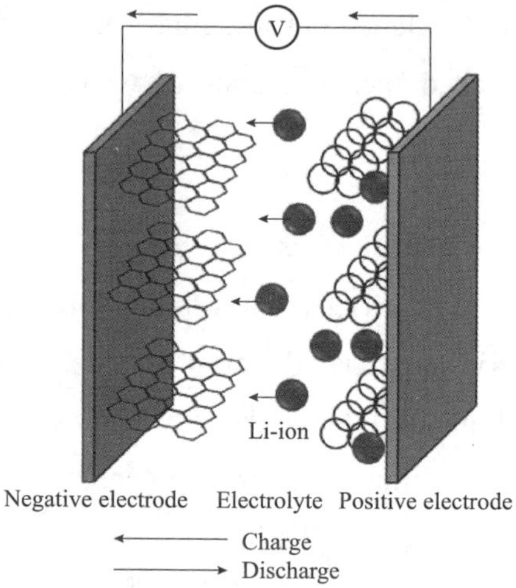

Figure 3　Principle operation a Li-ion cell

There are many different types of lithium-ion batteries and they are named after the active materials. One of the most common lithium-ion batteries is the lithium cobalt oxide ($LiCOO_2$) type often abbreviated as LCO. Cobalt is the main active material that gives the character to the battery. The cathode is of cobalt oxide while the anode is of graphite. The cathode has a layered structure and during discharge lithium ions move from the anode to the cathode. The ion flow is reversed during charging. The drawbacks of the LCO are the relative short life span, the low thermal stability and the limited specific power. Another lithium-ion cell uses lithium manganese oxide ($LiMn_2O_4$) as the cathode material. A three-dimensional **spinel** structure is formed that improves the ion flow in the electrode and accordingly the internal resistance is reduced. Also the spinel structure provides high thermal stability and enhanced safety. Fast charging and high-current discharging are possible. The lithium manganese batteries are applied as power tools, medical instruments and for hybrid and electric vehicles. Another lithium-ion concept has a cathode combination of nickel-manganese-cobalt (NMC). The combination is typically one-third of each Ni, Mn and Co. By this, the raw material cost can be reduced as less cobalt is used. There is a trend toward NMC blended lithium ion batteries as these can be built economically and achieve good performance as they enable high capacity and high power. Applications of NMC are E-bikes, medical devices electric vehicles and energy storage systems. Still another lithium-ion type battery is the lithium-nickel-cobalt-aluminum oxide ($LiNiCoAlO_2$) battery abbreviated NCA. It has similarities with the NMC but adding aluminum gives the chemistry greater stability. Lithium titanate ($Li_4Ti_5O_{12}$) is another li-ion battery where lithium titanate replaces the graphite in the anode and this material forms a spinel structure. The cathode can be LMO or NMC. It has comparatively low specific energy but is very safe. Sometimes this battery is abbreviated as LTO. Phosphate has also been discovered as a suitable cathode material for rechargeable lithium-ion batteries. In the Lithium-iron-phosphate ($LiFePo_4$) cell the anode is of graphite. The cell offers high current rating, long cycle life, good thermal stability and high safety.

For a lithium-ion battery operating with $LiCoO_2$ as the positive electrode (cathode) and carbon as the negative electrode (anode), the overall or net reaction is:

$$LiCoO_2 + C \longleftrightarrow Li_{1-x}CoO_2 + Li_xC$$

In the lithium-ion polymer battery (LiPo), the electrolyte is a polymer matrix, gelled or plasticized by a liquid. The electrolyte is in solid state and the absence of free liquid provide these batteries to be more stable and less vulnerable to problems associated with overcharge, damage or abuse. However, the electrolyte cannot be made as thin as for the lithium-ion concept and thus the effective surface area of the electrodes and accordingly the power capacity are limited. The nominal voltage is 3.7V. The manufacturing cost is also high. The LiPos are commonly used in portable devices. They are deemed to be robust and as they are used they safely constitute great power sources.

5 Zinc–Air Batteries

In a zinc-air battery, zinc and oxygen work together to generate power. This battery is completely filled with zinc, which reacts with oxygen from the air when the seal is removed. Frequent applications of this battery type are in hearing aids and watches. The zinc-air battery is basically a primary cell (often in button or coin cell sizes) but rechargeable designs for high power applications (like traction) are possible by mechanically changing the electrodes. In general, the energy density is high but the power is low. As zinc is an inexpensive material, zinc-air batteries are commonly relatively cheap.

The positive electrode or the cathode is a porous unit made of carbon with air access. The atmospheric oxygen is reduced at this electrode. The negative electrode or the anode consists of zinc. The electrolyte is an aqueous solution of potassium hydroxide.

The air electrode is divided into a catalytic active layer and a porous diffusion layer which is permeable to oxygen as well as being hydrophobic. The oxygen partial pressure difference between the atmosphere and the interior of the cell is taken advantage of in the operation. The catalyst enables the reduction of oxygen to hydroxyl ions, which migrate from the cathode through the electrolyte to the anode to complete the redox reaction. During discharge the following reactions take place:

Negative electrode, anode: $Zn + 4OH^- \longrightarrow Zn(OH)_4^{2-} + 2e^-$

$$Zn(OH)_4^{2-} \longrightarrow ZnO + H_2O + 2OH^-$$

Positive electrode, cathode: $O_2 + 2H_2O + e^- \longrightarrow 4OH^-$

Overall reaction: $2Zn + O_2 \longrightarrow 2ZnO$

The cell voltage for the chemistry is theoretically 1.65V. However, most of the designs are optimized for a potential in the range of 1.2~1.4V in order to achieve longer life times.

Reversing the processes above to recharge the zinc-air battery is associated with difficulties. Zinc, as many other metals, suffers from unequal deposition on the anode side during charging. Spiky dendrite crystals are formed and these can grow large to short out the cell. At the cathode it is difficult to electrically reverse the reaction to liberate oxygen from the hydroxyl ions and water. It requires catalysts of expensive metals like platinum. However, so-called mechanical recharging has been developed. Then the zinc oxide discharge product at the anode, the entire anode and the electrolyte can be replaced. Fresh zinc powder and electrolyte can be combined into the consistency of a slurry which can be pumped into the battery. The zinc oxide can be removed and converted back to zinc metal at an alternate processing site. This eliminates the problem with dendrite formation at the anode. Mechanical recharging is attractive as it can provide a kind of fuel tank refilling similar to gasoline using vehicles. By removing the spent zinc oxide and replacing it with fresh metallic zinc, it is possible to recharge the battery quickly.

Selected from "*Hydrogen, Batteries and Fuel Cells*", Bengt Sunde'n, Elsevier, 2019.

Notes

Capacity: *noun.* (1) the maximum amount that something can contain. (2) the amount that something can produce. (3) a specified role or position.

Energy density: *noun.* the amount of energy stored in a given system, substance, or region of space per unit volume.

Electrolyte: *noun.* a liquid or gel that contains ions and can be decomposed by electrolysis, e.g., that present in a battery.

Insertion: *noun.* (1) the action of inserting something. (2) an amendment or addition inserted in a text. (3) the manner or place of attachment of an organ.

Spinel: *noun.* a hard glassy mineral occurring as octahedral crystals of variable color and consisting chiefly of magnesium and aluminum oxides.

Reverse: *verb.* move backward. *adjective.* going in or turned toward the direction opposite to that previously stated. *noun.* (1) a complete change of direction or action. (2) the opposite or contrary to that previously stated.

Questions

(1) Batteries are devices that transform chemical energy into electricity. What are the main components of battery? Please describe how batteries work.

(2) Why the electrode be manufactured in electrically and ionically conducting materials?

(3) What are the main tasks of electrolyte and the separators?

(4) The single-use batteries are widely available and used in our daily life. However, they must be disposed after being fully discharged. Please explain why the disposed batteries should be collected and recycled.

(5) The rechargeable batteries are built to be recharged many times. They generated less waste than single-use battery. However they are more expensive. What electronic devices do you know have rechargeable batteries? Please give some examples.

(6) Please describe the working cycle of dischargeable lead batteries.

(7) What are the reactions at both anode and cathode of lead batteries?

(8) What main elements do the NiMH cell have? What are the reactions of NiMH batteries?

(9) Lithium-ion batteries are today the most frequently used rechargeable batteries and appear in laptops, cell phones, digital cameras and hybrid and electric vehicles. What are the advantages and drawbacks of Lithium cobalt oxide type batteries?

(10) Zinc-air battery have been frequently used in hearing aids and watches. Please describe how Zinc-air battery work.

Part B Reading for Innovation

Applications of Batteries

1 Introduction

Battery design is rapidly evolving for both consumer and industrial applications. Consumer devices are not designed to operate in industrial environments. Billions of batteries are manufactured globally each year, the vast majority of which are used in consumer devices such as flashlights, cameras, laptops and cell phones. These consumer applications require both rechargeable and primary (non-rechargeable) batteries. Batteries that are used in handheld devices are usually easily recharged or easily replaced, and are designed to operate at moderate temperatures.

Devices such as cell phones are usually replaced every two years as part of a contract upgrade, and laptop/tablet devices quickly become **obsolete**. As a result, most consumer grade rechargeable batteries are only required to operate for a few years with approximately 500 full recharge cycles. Other consumer products such as flashlights, remote controllers and toys typically use inexpensive primary lithium batteries that are also designed to operate within a temperature range (0 to 40℃) with an expected lifespan of $2e^3$ years. Unlike consumer applications, industrial devices are often used in remote, hard-to-reach locations where the device needs to be self-powered and battery replacement and recharging is difficult or impossible. Industrial applications have unique power needs and the choice of batteries is important. In industrial applications, the focus is more on durability and reliability in contrast to consumer products which have strong demands on high energy density to enable slim and elegant designs.

Accordingly, industrial batteries are bulkier than those being used in consumer products but on the other hand achieve a longer service life. Certain industrial applications must withstand extreme temperatures ranging from –55 to 85℃, with cold chain applications going down to –80℃, and high temperature applications reaching as high as 150℃. Concerning vehicles, most of the news concerning electrification is about passenger cars. However, there is also a significant shift toward electric vehicles in transport and haulage. It should be noted that light-duty trucks represent almost as much of the market as passenger cars, and thus emissions from those vehicles deserve at least as much attention as the emissions from passenger cars. However, the real benefits for the environment will come from replacing the diesel engines in medium and high-duty trucks with clean electric power. Trucks are available with various hybrid and full electric drive options. Fork lift trucks and mine trains have been used in industry for many years but suffered limitations due to the technology and the drive components being used. Advances in the battery technology are finding their way into heavy goods vehicles (HGVs) with good results.

2 Electrical Vehicles

Electric vehicles have different degrees of **electrification**, i.e., a certain amount of the energy consumption comes from electricity and the rest from an internal combustion engine. An all-electric vehicle (EV) is the simplest type among the electric vehicles as the battery is the single source of energy. This type is also called battery electric vehicle (BEV) or sometimes pure electric vehicle (PEV). The battery of a BEV acts like the fuel tank of a conventional vehicle, i.e., it can be filled up and the vehicle runs until the tank is empty. However, for a BEV the refueling or recharging time might be long. The state-of-charge SOC is an indication of the charging status of the battery and 100% means a fully charge battery.

A hybrid electric vehicle (HEV) is a vehicle having two or more types of power sources and at least one is an electric one. Brake energy is regenerated and stored in the battery and this is an important feature of HEVs concerning the total energy efficiency. The battery of an HEV can only be charged by recuperating the brake energy or utilizing the internal combustion engine, i.e., no external charging procedure is required. From the user point of view, the HEV is much like a conventional vehicle.

The HEVs come with different degrees of electrification and these are called micro-HEV, mild-HEV and strong-HEV. Only for strong-HEVs, a limited all electric driving range is possible but the distance depends on the battery capacity. **Plug–in** hybrid electric vehicles (PHEVs) combine the advantages of the BEV and HEV. The battery can be charged externally and the all-electric driving range will be longer than for the strong HEVs and the overall driving range will be longer than for the BEVs. The capacity of the battery is important for the all-electric driving range. The ability to recharge the battery externally by grid electricity, regenerative brake energy or by the internal combustion engine creates opportunities to optimize the energy consumption and accordingly the selection of battery.

Below the focus will be on BEVs. Several cars using Lithium ion cells are now on the market, e.g., Nissan Leaf EV, Chevrolet Volt, Renault Fluence, Subaru G4e, BYD E6, Tesla Model S, Honda Fit EV and RMW i3. Table 1 provides data for some BEV cars. Figure 1, shows conjectured picture of a BEV at the charging station.

Table 1. Characteristics of some BEV cars

Car	Capacity/(kW·h)	Weight/kg	Energy density of pack/(W·h/kg)	Type of cell
Nissan Leaf	24	294	81.6	Pouch Cell
Renault Zoe	22	290	75.9	Pouch Cell
BMW i3	18.8	230	81.7	Prismatic Cell
Tesla Model S	85	600	141.7	Cylinrical

Figure 1　A BEV at a charging station

3　Battery Types for Electric Vehicles

Battery electric vehicles (BEV) have the advantage of generating no greenhouse gases or pollutants, i.e., zero emissions. Hence, BEVs contribute to a cleaner air and an improved environment. There are many different models of commercially successful

BEVs from the more economical Nissan Leaf to the more expensive Tesla Model S. As there have been and still are great investments in support of infrastructure, the BEVs have become more viable and feasible options in the automotive market for consumers. There are several different types of batteries that have been implemented in battery vehicles.

(1) Lead Acid Batteries and Nickel Metal Hydride Batteries (NiMH)

Both lead acid batteries and nickel metal hydride batteries are regarded as mature battery technologies. These types of batteries were originally used in early electric vehicles, e.g., General Motor's EV1. However, currently they are considered to be obsolete with regards to their usage as the main source of energy storage in BEVs. Lead acid batteries have been used for many years in conventional combustion engines driven vehicles and are relatively in expensive. However, this type of battery has a poor specific energy about 34 W·h/kg. NiMH batteries are considered to be superior, compared to lead acid batteries, as they can have a specific energy as high as 68 W·h/kg. This allows electric vehicles utilizing NiMH batteries to be significantly lighter leading to reduced energy cost for propelling the BEVs. Similarly, NIMH batteries also have greater energy density than lead acid batteries, which will allow the battery system to be contained within a smaller space. However, NIMH batteries also have some drawbacks, e.g., lower charging efficiencies than other batteries. Self-discharge is another issue for NiMH batteries. This might be up to 12.5% per day at normal room temperature conditions and this is exacerbated when the batteries are in high temperature environment. This matter makes the NiMH batteries less ideal for hotter environments. In addition, there have been legal controversy regarding large format NiMH batteries, which has affected the use of NiMH batteries in battery electric vehicles. Lead acid batteries are commonly used in wheelchairs, scooters, forklifts and golf cars.

(2) Lithium-Ion Batteries

Lithium-ion batteries are now considered to be the present standard for modern BEVs. There are many types of Li-ion batteries that each has different characteristics but the vehicle manufacturers keep focus on variants having excellent longevity. Compared to other mature battery technologies, the Li-ion batteries offer many benefits. Among these, the excellent specific energy, about 140W·h/kg, and the energy density, make this battery type ideal for BEVs. Li-ion batteries are also excellent in retaining energy, with a self-discharge rate of 5% per month. This is an order of magnitude less than that for NiMH batteries. Nevertheless, Li-ion batteries also have some drawbacks. Compared to other technologies the Li-ion one is expensive. Some safety concerns regarding the overcharging and overheating exist. Li-ion batteries can experience thermal runaway, which can trigger vehicle fires or explosions. So for instance, the Tesla Model S, which uses Li-ion batteries, had infamously caught on fire due to issues with **fluctuating** charging or damage to the battery. However, great efforts have been made to improve the safety of vehicles using Li-ion batteries.

Batteries for traction. For wheelchairs, scooters, golf cars and forklifts, most commonly lead acid batteries are used for powering. The heavier the wheeled application becomes, the less suitable a battery will be. However, this does not prevent engineers and industries to develop large battery systems to replace the polluting internal combustion engines. The automatic guided vehicle (AGV) system as ship ports is an example. AGVs run almost for 24h per day so there is no time available for long charging intervals. Batteries of lithium-ion type solve this partly by replacing heavy lead acid batteries and provide lighter batteries which also can be recharged quickly. For large traction systems, no economical battery solution exists yet.

(3) Estimation of the Weight of a Long Haulage Truck

A typical truck of 40000kg weight drives about 200000km per year. The energy consumption can be estimated as 600000kW·h per year. The daily travel might be 650 km. The energy consumption per day is then close to 2000kW·h. The electric propulsion reduces the energy by 50%. If the battery system has a cycling capacity of 50W·h/kg, the weight of the battery system becomes 20000kg. This means that the truck has to carry a huge and heavy battery if a battery solution should prevail for the power train.

From the above one can learn that the heavier the wheeled application is, the less suitable the battery becomes. In addition, the high price needs to be cut down, the limited driving range must be extended without a too bulky solution and the charging time has to be reduced.

The needs to carry very heavy loads across long distances make long-haul trucking one of the hardest vehicle segments to electrify. At present, the three types of all-electric propulsion options in the long-range trucking market are batteries, hydrogen fuel cells and overhead catenary systems (power supply by an overhead wire system). Each of these has both barriers and options.

Catenary system. Trucks powered by overhead wires have been used in mining for several years but this is a specific application with a limited circuit and low speeds. The principle of overhead power feeds to trucks on short sections of a highway is piloted in several countries now. The system is based on trucks with a hybrid electric drive which can switch from electric to diesel or natural gas power. The purpose of the pilots is to evaluate the effectiveness of a catenary system. The present application is short hauls of heavy goods. For a battery electric truck, the battery is allowed to be recharged when driving on such a way. Figure 2 shows a principle sketch of an electric truck being fed by overhead power.

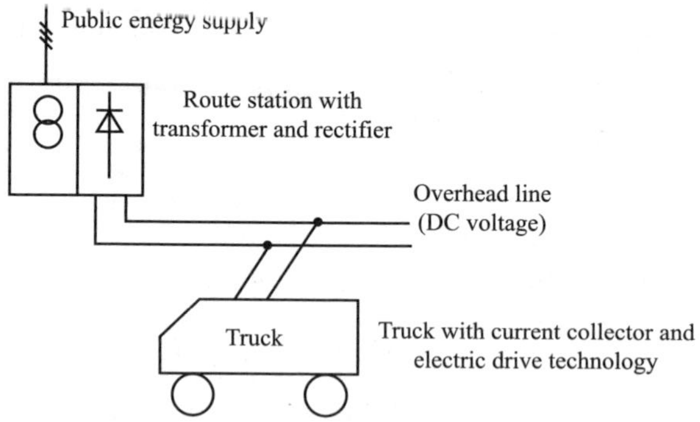

Figure 2　Principle sketch of an electric truck with overhead power feed

Hybrid systems. Hybrid powertrain systems have been slow to migrate into the HGV market but some models are already in use. Hybrids are used typically for short haul trips with a large number of start-stop-actions which facilitate regenerative charging Applications include urban delivery and utility vehicles with single shift operation allowing slow battery recharging at off-shift or off-peak times.

Battery electric goods vehicle. The difference between the battery-powered passenger vehicle and the goods vehicle is the amount of space available for the battery. A goods vehicle can accommodate a larger battery although this may significantly reduce the hauling capability and payload. An ultimate aim is an autonomous electric drive vehicle which can haul heavy goods over long distances without a need for recharging. Although this goal is being approached, it is far from realized and although heavy-duty BEVs are available, their range is limited by the size and mass of the battery as well as by the absence of suitable recharge stations along the route. Factors limiting the use of BEVs to urban distribution applications are battery weight and recharge times. At the heart of the development is the Li-ion battery, in various forms, which offers high energy and power density than lead-acid batteries. Vehicles may use a single electric motor or a system of electric motors on each wheel or on each axle. This allows control of adverse solutions such as jack-knifing.

(4) Batteries for Commercial Vehicles

Larger vehicles like motor coaches, semi-trucks and buses need a high-quality performance battery equipped to sustain long trips, frequent stops and slowing down. To power commercial vehicles, three main types of heavy-duty batteries exist, namely lithium iron phosphate ($LiFePO_4$), GEL, and absorbed glass mat (AGM). All these three types are considered as deep cycle batteries, which are intended to be used up to the maximum capacity. Deep cycle batteries can withstand many charge cycles and are perfect for use during extended periods of time. It should be noted that starter batteries provide surge power needed to start an engine while deep cycle batteries provide continuous power.

Lithium-iron-**phosphate** batteries or flooded lead-acid batteries are cost effective power solutions that need to remain wet. Electrolyte fluid floods the top of the battery plates. Based on the discharge rate for each cycle, they have a life span of about 200~300 cycles. This low life span is largely due to internal grid corrosion and loss of active material. Sealed lead acid batteries, also known as valve regulated lead acid batteries (VRLA) include gel and absorbed glass mat varieties. These types are maintenance free and do not need extra care such as adding water or cleaning. Gel and

AGM are ideal for their high performance, no gas emission and longer charge life.

Gel batteries use a silica-type gel, which suspends the electrolyte absorbed glass mat batteries suspend the electrolytes through glass mats which enables faster charging. Gel and AGM have many similar features but AGM has a higher burst of amperes. On the other hand, gel batteries have a slower discharge rate.

A 12V AGM battery has some similarities to a standard lead acid battery, but differs in the sense that it holds the electrolytes in the glass mats, rather than flooded plates. Woven on the mat is thin glass fibers, as this enhances the surface area. This allows them to hold enough electrolytes on the cells for the lifespan of the battery. These fibers are not affected or absorbed by the acid of the electrolytes.

4 Batteries for Aviation

Batteries on board an aircraft serve to feed **navigation** and emergency systems when the auxiliary power unit (APU) is off or during an emergency in flight. The batteries provide power for barking, ground operation and start up of the APU. In case of engine failure, the batteries must be able to supply energy from 30min to 3h. In addition, an aircraft must also have enough battery power to enable a safe landing. During flight, the electrical power is supplied by generators. Most commercial jet airplanes use flooded nickel-cadmium batteries. During the start-up phase of a large aircraft, first the APU (a small turbine engine at the tail section) is spooled. The spooling speed of the APU must be sufficiently high to attain the compression for self-sustained ignition. Once running, an air compressor or hydraulic pump jumpstarts the large jet engine one-by-one. Smaller aircrafts commonly use lead acid batteries. These are heavier than NiCd batteries but require less maintenance. Modern jet fighters spool the jet engines by Li-ion batteries, e.g., the Boeing 787 Dreamliner. Airbus 350 offers either chemistry optionally. As the on-board functions of an airliner move from hydraulic to electrical, larger batteries are needed. Here Li-ion batteries serve the demand better as they provide higher energy density than NiCd and lead-acid batteries.

NiCd batteries maintain high durability and reliable service but they need maintenance and the memory problem must be eliminated.

Although an aircraft may carry many different batteries onboard, their purpose is to start the engines and provide backup power when the engines are off. It is likely that large aircrafts will continue to fly on fossil fuels or biofuels as batteries are not yet practical for propulsion. However, small battery-powered airplanes are used for pilot training and to fly short hops but still under development.

Weight and reliability off aging batteries remain major concerns.

5 Batteries for Aerospace

Early satellites were using NiCd batteries and this led to the discovery of the phenomenon memory. The battery followed a routine discharge schedule but when more energy was demanded, the battery could remember. The voltage would drop as if to protest against an unwanted overtime. NiCd batteries were replaced by nickel-hydrogen as a battery with an exceptional long service life. Trials were made to introduce this battery for commercial use but high price and large size spoiled the market acceptance.

For satellites, Li-ion batteries are chosen. These provide low weight, easy charging, durability and good cycling behavior. The self-discharge is low and maintenance is almost not needed.

The Mars curiosity Rover uses specially designed lithium nickel oxide cells (LiNiCo) in a formation of eight cells in series and two in parallel. Partial charging and discharging are used to stretch longevity. Under this condition, the life span is four years and approximately 700sol (sol the duration of a solar day on Mars). The 34A·h cells, of which two are in parallel, have a maximum discharge rate of 0.55C.

NASA has a requirement on Li-ion batteries to last for 7 years and 37000 cycles with a DoD (depth of discharge) of 40e60%. NASA laboratories reveal that the end-of-life (EOL) is connected with the growth of the SEI (solid-electrolyte-interface)-layer on the anode, loss of cathode material, loss of conductive path, plating of metallic lithium and electrolyte oxidation. Large 140A·h Li-ion cells are in development that promise to last up to 18 years.

Selected from "*Hydrogen, Batteries and Fuel Cells*", Bengt Sunde'n, Elsevier, 2019.

Important Words and Phrases

Obsolete: *adjective*. (1) no longer produced or used; out of date. (2) (of a part or characteristic of an organism) less developed than formerly or in a related species; rudimentary; vestigial. *verb*. cause (a product or idea) to be or become obsolete by replacing it with something new.

Electrification: *noun*. the action or process of charging something with electricity.

Plug–in: *adjective*. able to be connected by means of a plug.

Fluctuate: *verb*. rise and fall irregularly in number or amount.

Phosphate: *noun*. a salt or ester of phosphoric acid.

Navigation: *noun*. (1) the process or activity of accurately ascertaining one’s position and planning and following a route. (2) the passage of ships.

Part C Reading for Fun

Lithium Batteries Recycling

1 Introduction

There can be several different recycling strategies for different products or product groups. Typically, they can be placed somewhere in a coordinate system, as shown in Figure 1.

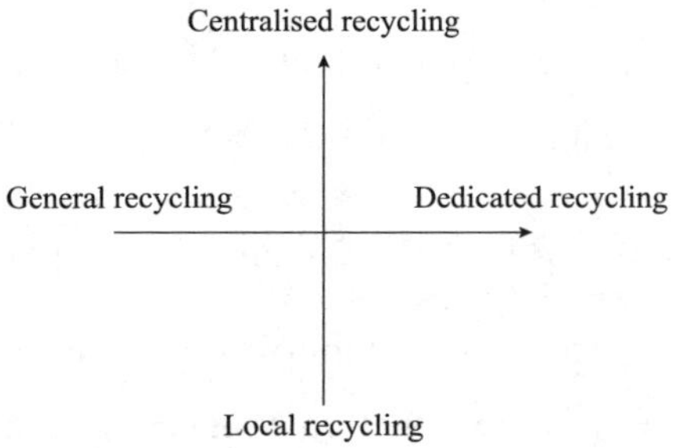

Figure 1 Different recycling strategies

Dedicated recycling can be made very specific, thus generating a minimal amount of secondary waste and possibly resulting in high purities. On the other hand, it is sensitive to changes in the product flow and the processes are typically also complicated and require special competence.

General recycling is the typical fragmentation plant. Almost anything can go in and be divided into smaller parts, which are then separated into more or less pure streams. This kind of process is very robust and often rather simple as a principle (although not always since some sorting techniques can be complicated). Except for the problem in obtaining high purity streams there are often some secondary waste streams that have to be landfilled or otherwise dealt with.

Local recycling is often placed close to the actual manufacturing site. Therefore it can handle not only products but also manufacturing waste and discarded products. The resulting product streams can then go directly into the manufacturing process again. Since many of these are needed for different manufacturing sites there will be a need for a large number of competent workers and specialist knowledge on site.

Centralized recycling profits from the large-scale operation. A small number of facilities can serve large areas. Transport is naturally an issue but the gain in large product flows typically makes these more profitable. The products often go onto the open market rather than directly to the manufacturer.

In the case of batteries, it is possible to consider several of these alternatives. For small and portable batteries general recycling is possibly the best, due to differences in size and chemistries making efficient sorting of them difficult. When it comes to car batteries it may be better to have centralized, dedicated recycling where, e.g., a plant can be built up to handle a specific type of battery but the collection area may involve several countries. We will discuss some techniques and practices in relation to the recycling of lithium (and other metals) from different types of lithium batteries (LIBs).

2 Characterization of Spent LiBs Waste

LiBs composition varies from one technology to another, depending on their applications. For instance, it is clear that components and materials used in the next generation of LiBs for electrical vehicles will be quite different from those presently developed for mobile phones or laptops due to health and safety, economic, and technological reasons.

In general, spent LiBs consist of a steel and aluminum casing, printed circuit boards, plastic covers, copper and aluminum foils, separators, active material, carbon, organic binders, organic solvents, salts, additives etc. According to the European Battery Recycling Association (EBRA), spent portable LiBs are typically composed of Al foil (15%~25%), carbon (0.1%~1%), Cu foil (5%~15%), diethylcarbonate (DEC, 1%~10%), ethylene carbonate (EC, 1%~10%), methylethyl carbonate (MEC, 1%~10%), $LiPF_6$ (1%~5%), graphite (10%~30%), $LiCoO_2$ (25%~45%), PVDF (0.5%~2%), steel, nickel, and polymers. More precisely, the material safety data sheet for portable

LiBs (3.7V) made by Varta introduces the following material composition: carbon (10%~30%), $LiCoO_2$ (20%~50%), electrolyte (10%~20%), Cu foil (2%~10%), Al foil (2%~10%), steel (50%~80%), polymers (2%~10%). Table 1 shows some examples of the material composition of spent LiBs reported in the literature.

Table 1. Typical Material Composition of Spent LiBs

Component /(wt%)	Source 1*	Source 2*	Source 3*
Anode material	39.1	16	—
Cathode material	17.8	27.5	—
Separator	2.5	—	5.2
Electrolyte	1.2	3.5	4.7
Cover	34.3	24.5	10.5
Polymers	—	14	22.9
Cu/Al ratio	—	14.5	—
Cu foil	—	—	8.9
Al foil	—	—	6.1
Black mass	—	—	39.1

*Check original article for details.

Researchers studied the characterization of spent portable LiBs. In this work, approximately 15kg of spent materials were obtained after manual separation of LiBs. After crushing, milling, and magnetic separation, two fractions with grain sizes of 0~1 mm and 1~3 mm, respectively, were obtained. Solid samples were dissolved in aqua regia in a microwave oven at 210℃. ICP-AES analysis of the leaching solution is reported in Table 2 and the results are compared with those obtained in two other studies.

Table 2. Metal Composition of Spent LiBs after Direct Crushing

Element	Fraction 0~1nm*	Fraction 1~3nm*	Source 5*	Source 6*
Al	0.54	2.6	13.1	0.23
As	—	—	—	—
Cd	—	—	—	—
Co	26	22	23.3	22.7
Cu	1.9	12	12.2	0.29
Fe	0.56	0.092	1.8	6.15
Hg	—	—	—	—

(continued)

Element	Fraction 0~1nm*	Fraction 1~3nm*	Source 5*	Source 6*
K	1.1	1.4	—	—
Li	3.2	2.7	2.7	2.34
Mg	—	—	—	—
Mn	0.91	0.54	0.11	—
Na	0.33	0.3	—	—
Ni	11	11	1.41	0.78
Pb	—	—	—	—
Sb	—	—	—	—
Sn	1	0.44	—	—
Ti	—	0.91	—	—
V	—	—	—	—
Zn	0.09	0.91	—	—
Others	<53	<45	—	—

*Check original article for details.

It appears that cobalt content is approximately 23% and lithium content is 2%~3% in these samples after direct crushing. Table 3 shows anode and cathode compositions from manually dismantled spent LiBs as reported in several studies:

Table 3. Metal Composition of Spent Electrodes from LiBs after Manual Dismantling

	Element/ (wt%)	Source 7	Source 8	Source 9	Source 10
Cathode	Li	6.54	5	4.4	3.14
	Co	52.9	43.3	53.8	47.96
	C	—	—	—	—
	Al	11.03	10.2	0.8	12.2
	Ni	1.14	—	0.97	—
	Mn	—	—	—	—
	Cu	—	0.7	—	—
Anode	C	—	—	—	—
	Li	—	0.5	—	0.79
	Co	—	0.047	—	3.22
	Cu	—	40.7	—	52.64
	Al	—	0.054	—	1.93

According to these results, cobalt and lithium contents are approximately 40%~50% and 3%~6.5% in the spent cathode materials, respectively, whereas 12% aluminum comes from current collectors at the cathode. The anode consists mostly of copper (40%~50%, current collector) and carbon (active material). X-ray diffraction (XRD) analysis of spent materials from LiBs show that these wastes usually contain $LiCoO_2$, C, Co_3O_4, Al, and Cu.

In a recent study, researchers characterized materials after applying chemical and mineral processes to spent LiBs. Discharged batteries were crushed in shear and impact crushers. Samples were sieved in order to obtain eight groups of different particle sizes. Before chemical and mineral characterizations, samples were frozen at −196℃ and then ground.

Battery waste is a significantly heterogeneous and poly-component material. This waste contains not only inorganic compounds, but also organic components. The chemical composition of the fractions is highly dependent on the mechanical pretreatment. Nowadays the main goal of recycling processes is to recover valuable components such as cobalt, copper, nickel, (lithium), etc., which are concentrated in the black mass with the exception of copper. Thus the efforts of the mechanical pretreatment are focused on the recovery of the black mass and the separation of the particular components (aluminum foils, copper foils, plastic and steel parts, etc.).

3 Recycling of Spent Portable LiBs

In general there are four types of recycling technologies: mechanical treatment, hydrometallurgical treatment, combination of thermal pretreatment and hydrometallurgical methods, or pyrometallurgical treatment.

Mechanical treatments include crushing and physical separation methods to separate the particular components and recover the black mass, which contains the valuable metals (mostly cobalt, nickel, manganese, lithium, etc.). Hydrometallurgical technologies implement mechanical pretreatment and metals recovery from the black mass by means of leaching, precipitation, solvent extraction, and ion exchange resins. Pyrometallurgical technologies are focused on the production of the metal alloys. Spent LiBs are usually processed without any mechanical pretreatment. The material recycling efficiency of pyrometallurgical processes is lower than that of hydrometallurgical processes. Plastic covers and all organic compounds are incinerated.

Metals such as aluminum, manganese, and lithium are lost in the slag.

The main focus of the recycling activities is to recycle cobalt since this metal exhibits high economic value (32.5$/kg), but the recovery of less economically valuable metals (such as lithium) is also becoming an interesting option for recycling companies.

Lithium carbonate is the precursor for the active battery material production. Today, depending on the purity of lithium carbonate, the price of the lithium carbonate is 6~15.5$/kg.Approximately 120000 tons of lithium carbonate are produced per year. More than 28%~35% of that lithium production is used in the battery industry; 26%~28% is used for the frits and glass production. Lithium is also needed for the production of lubricating greases, the air conditioning industry, metallurgy, medical areas, etc.

(1) Dismantling and Mechanical Pretreatment

Mechanical pretreatment of spent LiBs consists of two main processes: disintegration of batteries and separation of the particular fractions. Disintegration applies processes such as crushing, milling, and shearing, etc. Separation techniques are intended to separate the materials based on their difference of physical properties such as density, conductivity, magnetic properties, etc. Figure 2 shows the general flow sheet usually implemented for the mechanical pretreatment.

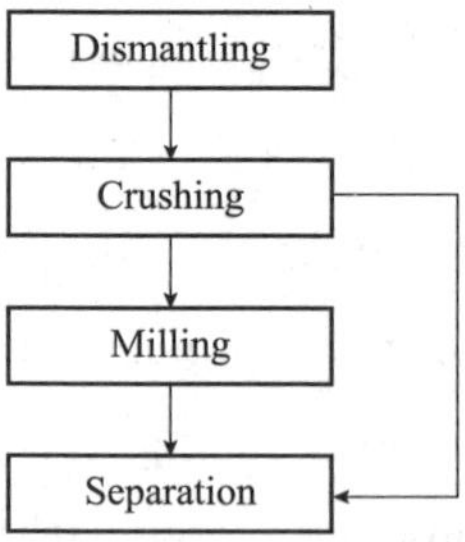

Figure 2. General flow sheet of mechanical pretreatment

Discharging Spent Batteries. For safety reasons spent batteries usually undergo a discharging step before performing manual dismantling to avoid short-circuits and sparks when the cells are not discharged which might cause the ignition of volatile organic compounds during the crushing process. This discharging step can be performed by different methods. The most common method is based on the immersion of the batteries in a salt solution or distilled water to deactivate the cells.

Dismantling of the Spent Batteries. While portable batteries are usually crushed directly without any previous special handling, car batteries require manual dismantling. Particular cells have to be removed from the battery. Plants for manual dismantling have been established, e.g., by Umicore in Germany and the United States. Car batteries are dismantled and battery cells are transported to the recycling plant.20 General flow sheet of manual dismantling is shown in Figure 3.

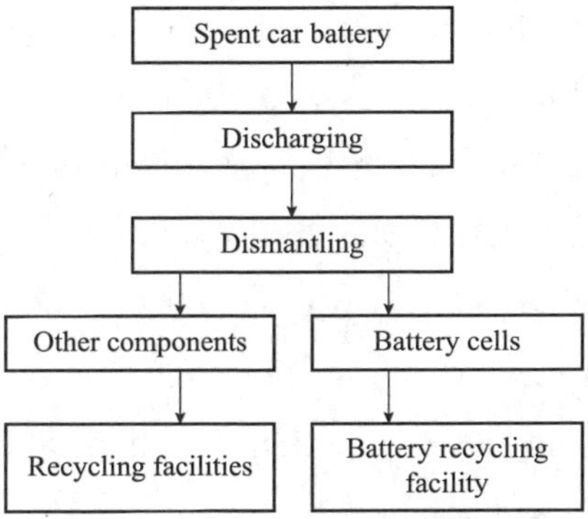

Figure 3. General flow sheet of manual dismantling

In general, spent batteries have to be discharged before dismantling. Afterward, particular components such as cables, cooling units, printed circuit boards are sent to the recycling facilities while battery packs are sent to the battery recycling plant. In order to separate the black mass from the current collector, the organic binder made of PVDF has to be dissolved. Organic solvents such as *N,N*-dimethylformamide(DMF), *N,N*-dimethylacetamide(DMAC), *N*-methylpyrrolidone(NMP), and dimethylsulfoxide (DMSO) can be used to dissolve PVDF. For instance, $LiCoO_2$ was separated from the current collect by dissolving PVDF with the three following solvent: DMAC>DMF>NMP. NMP was able to separate both $LiMn_{1/3}Ni_{1/3}Co_{1/3}O_2$ and $LiCoO_2$ from the current collectors at 40℃ after 15min and at 100℃ after 1h, respectively. The disadvantage of this process is the price of NMP which makes difficult to apply NMP for larger scale operations. Thus the use of proper mechanical pretreatment is necessary to achieve good and cheap separation of the black mass from the other components.

Mechanical Pretreatment of Spent LiBs. A series of mechanical processes involving crushing, sieving, magnetic separation, fine crushing, and classification are carried out to yield enriched particles of the black mass in sequence. Multistage crushing and sieving result in separation of the metal-bearing particles from the waste. A magnetic separation is used to remove pieces of steel casing. In order to eliminate small pieces of current collectors attaching to the particles of the black mass, a fine crushing can be used.

A separation process involving the air and electromagnetic separation was proposed. The crushed products of spent LiBs were divided by sieving into four fractions with particle sizes >2mm, <2mm and >0.5mm, <0.5mm and <0.075mm, and <0.075mm. The fraction with a particle size >2mm was separated by air separation. The aluminum casing and the separator were recovered by this method. Electrostatic separation was used for the fraction <2mm and >0.5mm and the aluminum and copper foils and separator were separated. The third fraction with particle size <0.5mm and >0.075mm was ground and dry sieving was applied to obtain better separation of black mass from the foils. After sieving, the fraction >0.075mm was treated by electrostatic separation together with the second fraction. Fraction <0.075mm can be treated by flotation to recover $LiCoO_2$ and graphite.

Electrodynamic separation was applied in the technology proposed. Spent LiBs were crushed with a two-rotor crusher and a hammer crusher, followed by sieving and thermal treatment to separate the black mass. Electrodynamic separation was used for material with a particle size greater than 1mm. Fractions of ferrous metals and nonferrous metals were obtained after the separation.

The black mass can be separated from the aluminum foil using ultrasound separation. This process consists of three steps: ① spent lithium batteries (LiBs) were crushed with a 12mm aperture screen; ② the smaller materials from the 12mm aperture were put into an ultrasonic washing machine with agitation equipment for 15min; and ③ the washed materials were put into a container with a 2mm aperture screen. Through this process, three types of products were obtained. The first product was a separator; the second one consisted of the steel casing, aluminum, and copper foils; and the third product was the black mass.

Air and magnetic separation are used to recycling LiB at the industrial scale by Akkuser (Finland). The hydrometallurgical plant, Recupyl (France) implements magnetic and density separation. Other companies such as Batrec and Retriev use mechanical pretreatment of spent LiBs as well.

(2) Hydrometallurgical Processes

A major effort has been made to develop the hydrometallurgical technologies for the recovery of metals from spent LiBs. The main focus concerned the recovery of cobalt, which is the most valuable component nowadays. On the other hand, hydrometallurgical treatment enables the recovery of lithium with a high purity, while pyrometallurgical treatment does not. In general, two approaches have been applied when studying the possibilities of spent LiBs recycling. Hydrometallurgical treatment or combined treatment, which consists of thermal pretreatment and hydrometallurgical treatment.

Hydrometallurgical treatment starts by discharging the spent portable LiBs, followed by dismantling, mechanical pretreatment, and separation stages. Processes such as leaching, solvent extraction, precipitation, and ion exchange are implemented to recover metals. Combined methods consist of mechanical and thermal pretreatment followed by hydrometallurgical treatment. Thermal pretreatment is used to remove organic compounds. Organic binders cause problems during the leaching and solideliquid separation stages. The presence of graphite was considered to cause a lower lithium extraction, because of lithium adsorption. A general flow sheet for combined methods is shown in Figure 4.

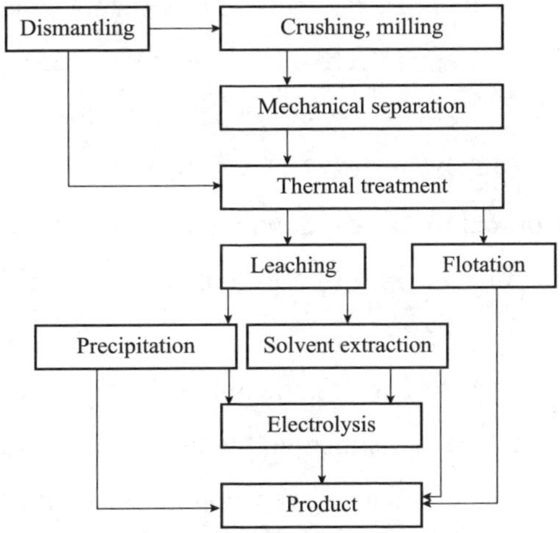

Figure 4. General flow sheet of the combined treatment

Thermal Pretreatment of Spent LiBs. In general, spent portable LiBs are recycled either by the hydrometallurgical or combined method at the laboratory and industrial scales. The hydrometallurgical method mostly includes mechanical pretreatment, leaching and extraction of the metal ions. The combined method usually includes mechanical and thermal pretreatments, extraction of metals by leaching and recovery of the metals. Thermal treatment is mainly used for removing organic compounds and graphite. Organic compounds (e.g., PVDF) are used as a binder and cause problems during the leaching and solid-liquid separation phases. The thermal pretreatment is usually performed in the presence of oxygen (incineration) or in the absence of oxygen (pyrolysis).

Recently pyrolysis has been applied as a thermal pretreatment method for organic component removal from spent LiBs. Pyrolysis is a thermal process performed without the presence of oxygen. Considering the chemical composition of the organic compounds in LiBs, pyrolysis seems to be a more environment-friendly process than incineration.

Leaching with Inorganic Acids. The concentrations and the nature of the leaching media, pH, solid to liquid phase ratio, temperature, etc., influence the leaching efficiency of spent materials. Inorganic acids such as H_2SO_4, HCl, or HNO_3 are usually used as the leaching medium.

Considering that $LiCoO_2$ is the most common active material, the reactions of active material with several acids are given by the following equations:

$$LiCoO_2 + 1.5H_2SO_4 = CoSO_4 + 0.5Li_2SO_4 + 0.25O_2 + 1.5H_2O$$

$$2LiCoO_2 + 6HCl = 2CoCl_2 + 2LiCl + 3H_2O + 0.5O_2$$

$$2LiCoO_2 + 6HNO_3 = 2Co(NO_3)_2 + 2LiNO_3 + 0.5O_2 + 3H_2O$$

Sulfuric, nitric, hydrochloric acids, aqua regia, ammonium hydroxide, and acetic acid have all been used for leaching. In addition, $NH_2OH \cdot HCl$ has been compared with H_2SO_3, and HCl.

High leaching efficiency of cobalt and lithium leaching has been obtained using inorganic acids, but as it has been mentioned, the highest efficiency was achieved using HCl. To increase the leaching efficiencies, reductive agents have been added to the acid in several studies.

Reductive Leaching. Since the chemical bonds between cobalt and oxygen are extremely strong, leaching of lithium cobalt oxide is difficult. Very often hydrogen peroxide has been used as a reduction agent. The addition of hydrogen peroxide generates oxygen that promotes cobalt oxide dissolution by reducing Co^{3+} to Co^{2+}.

The reactions of the active material with leaching medium and reduction agent proceed according to the following equations:

$$LiCoO_2 + 1.5H_2O_2 + 1.5H_2SO_4 = CoSO_4 + 0.5Li_2SO_4 + 3H_2O + O_2$$

$$2LiCoO_2 + H_2O_2 + 6HCl = 2CoCl_2 + 2LiCl + 4H_2O + O_2$$

$$2LiCoO_2 + H_2O_2 + 6HNO_3 = 2Co(NO_3)_2 + 2LiNO_3 + 4H_2O + O_2$$

Organic reducing agents can be also used to leach cathodes such as citric acid, malic acid, ascorbic acid, oxalic acid, and glucose. Glucose has been used as a reduction agent instead of hydrogen peroxide. The main disadvantage of those agents might be their price and negative impact on LCA (except Glucose). For the latter, the proposed reaction was:

$$24LiCoO_2 + 36H_2SO_4 + C_6H_{12}O_6 = 24CoSO_4 + 12Li_2SO_4 + 6CO_2 + 42H_2O$$

The black mass was leached using sulfuric acid in the presence of an excess of 50% of a reducing agent, such as glucose. The leaching efficiency of cobalt and lithium was 98%. Glucose was also used as a reduction agent. It was reported that the efficiency of glucose significantly depends on the time when it is added in solution.

When glucose is added at the beginning of the leaching, the leaching efficiency of cobalt is 60%, but when glucose is added after 2h of 80% of leaching efficiency of cobalt is achieved. It was reported that glucose can undergo different oxidation pathways during leaching. In particular the accumulation of a limiting intermediate in the gluconic pathway (preferentially occur when glucose is added at the beginning) could be the reason for the lower reducing ability observed under these conditions. Lithium extraction presented the same trend as cobalt. If glucose is added at the beginning, the extractive yield increases from 65% to 90% by stepping up the glucose excess. In contrast, when glucose is added after 2h, the same extraction yields, i.e., 92% are obtained with both glucose levels.

In general relatively high leaching efficiency of cobalt and lithium (80%~99%) has been obtained using HCl or H_2SO_4 with addition of hydrogen peroxide. The optimal concentration varies from 2 to 4M and optimal addition of hydrogen peroxide is from 1vol% to 6vol%. Leaching temperatures should be around 60~80℃ and a leaching time of 1h is sufficient for cobalt and lithium leaching.

Leaching with Organic Acids. Organic acids (citric acid, DL-malic acid) were also studied as a leaching agent for the recovery of lithium and cobalt from $LiCoO_2$. The leaching reactions are as follows:

$$7H_2C_2O_4 + 2LiCoO_2 = 2LiHC_2O_4 + 2Co(HC_2O_4)_2 + 4H_2O + 2CO_2$$

$$5H_2C_2O_4 + 2LiCoO_2 = 2LiHC_2O_4 + 2CoC_2O_4 + 4H_2O + 2CO_2$$

H_2O_2 is usually employed during leaching with organic acids in order to accelerate Co^{3+} reduction into Co^{2+} and to improve the leaching efficiency of cobalt occurring according to the following equations:

$$3H_2C_2O_4 + LiCoO_2 + 1.5H_2O_2 = LiHC_2O_4 + Co(HC_2O_4)_2 + 3H_2O + O_2$$

$$3H_2C_2O_4 + 2LiCoO_2 + H_2O_2 = Li_2C_2O_4 + 2CoC_2O_4 + 4H_2O + O_2$$

$$2H_2C_2O_4 + CoO = Co(HC_2O_4)_2 + H_2O$$

$$H_2C_2O_4 + CoO = CoC_2O_4 + H_2O$$

Bioleaching. Biometallurgical processes have been applied to recover metals from primary and secondary resources. The advantages of bio methods are sufficiently their low costs and low operational demands.

The chemolithotrophic and acidophilic bacteria *A. ferrooxidans*, which utilizes elemental sulfur and ferrous ion as energy source was used to produce metabolites such as sulfuric acids and ferric ion. The metabolites help dissolve metals from spent batteries. Bioleaching of cobalt was found to be faster than lithium. It was discovered that higher Fe^{2+} concentrations resulted in a decrease of the dissolution yield due to the coprecipitation of Fe^{3+} with the metals in the residues. A higher solid to liquid ratio also affected the metal dissolution by arresting the cell growth due to an increase of metal concentration in the waste sample. Bioleaching of $LiCoO_2$ at pH 2.5 with an S/L ratio of 5：1 in the presence of A. Ferrooxidans, iron, and sulfur (1% elemental sulfur and 3g/L Fe^{3+}) leads to the dissolution of 65% cobalt and 9% lithium, while only 20% of cobalt and 5% of lithium can be leached under the same conditions in the absence of bacteria. Researchers reported that the mechanisms involved in the bioleaching of $LiCoO_2$ electrodes are different depending on the nature of the media and the metal. For instance, bioleaching of lithium occurs due to the biogeneration of sulfuric acid, whereas the dissolution of cobalt is due to acid leaching and redox reactions involving iron and sulfur. Indeed, in the presence of iron and sulfur, H_2SO_4 and Fe^{3+} are generated resulting in direct acid dissolution of Co^{2+} and redox reactions leading to the formation of Fe^{2+}, which subsequently promotes the reductive dissolution of insoluble Co^{3+}.

Solvent Extraction and Precipitation. Bis(2-ethylhexyl) phosphoric acid (D2EHPA) is the most versatile extractant of phosphoric acids and has been used commercially for the extraction of many metal ions. Some particular advantages of the use of D2EHPA in solvent extraction processing are its chemical stability, generally good kinetics of extraction, good loading and stripping characteristics, low solubility in aqueous phase, and its availability in commercial quantities. From sulfate solution the order of extraction as a function of pH is: $Fe^{3+} < Zn^{2+} < Cu^{2+} < Co^{2+} < Ni^{2+} < Mn^{2+} < Mg^{2+} < Ca^{2+}$.

Production of New Material by Recycling. Several works have focused their research on the synthesis of new cathodes from materials produced from recycling. For instance, Researchers studied synthesis of $LiCoO_2$ from black mass of spent LiBs. Black mass was thermally pretreated at 120℃ for 12h. Material was milled and heated to 450℃ for 2h. The temperature was then increased to 600℃ for 5h. Material containing $LiCoO_2$, Co_3O_4, and $LiCO_3$ was mixed with $LiCO_3$ and calcinated at 850℃ for 12h. The quality of the material produced was comparable to primary material used in the production plant. Researchers prepared composite based on Li/Co ferrite. Spent LiBs were dismantled and cathode material was thermally pretreated between 150 and 500℃ for 1h. The black mass was mixed with Fe_2O_3 and calcinated at 900, 1000, and 1100℃ for 1h. Final product, $Li_{0.5}Fe_{2.5}O_4/CoFe_2O_4$, was produced at 1000℃ after 4h.

Researchers studied cobalt recovery from spent LiBs as Co_3O_4. Spent batteries were dismantled, dried, crushed, and pressed. Sulfuric acid (2M) and addition of 6vol% hydrogen peroxide was used as the leaching medium. Precipitation agent Na_2S was added to the solution after 30min to precipitate copper. After filtration 1.5M $H_2C_2O_4$ was added to the solution to precipitate cobalt. Precipitate $CoC_2O_4 \cdot 2H_2O$ was calcinated at temperatures from 150 to 600℃ for 2h. Approximately 99.9% of the copper was precipitated when a 3 : 1 ratio of Na_2S : Cu was used. More than 98% of the cobalt was recovered after precipitation with oxalic acid.

Selected from "*Lithium Process Chemistry: Resources, Extraction, Batteries and Recycling*", Alexandre Chagnes and Jolanta Swiatowska, Elsevier, 2015.

Basic Terminology in Batteries

1 Vehicle and Industry Terms

To begin simply, the abbreviation EV generally refers to an Electric Vehicle (EV), a vehicle that is fully electrified with no internal combustion engine (ICE) whatsoever. All of the power that is provided is done through the electric battery, powering one or more electric motors to provide propulsion and power to all of the other systems on the vehicle. These are also sometimes referred to as battery electric vehicles (BEVs), both terms are used interchangeably. However, the same abbreviation EV is also used to mean Electrified Vehicles (EVs), which refers to vehicles with all forms of electric power support including microhybrids, hybrid electric vehicles (HEVs), plug-in hybrid electric vehicles (PHEVs), and battery electric vehicles (BEVs). For the purposes of this book, I will use EV to mean only electric vehicles.

A PHEV is a vehicle that has both an electric propulsion system, generally ranging from 10 to about 40 miles of pure electric drive range, along with a combustion engine. This type of hybrid is often a parallel hybrid. In simplest terms, this means that the engine works in parallel with the electric motor. In this configuration, the electric motor is often "sandwiched" between the engine and the transmission. The benefit of the PHEV configuration is that it offers the full driving range that a comparable ICE vehicle would have—usually between 350 and 500 miles depending on the vehicle (Figure 1).

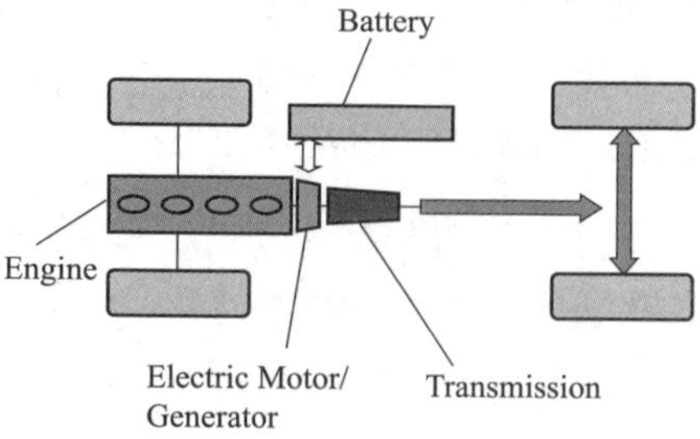

Figure 1 Parallel hybrid configuration

Similarly, the Extended Range Electric Vehicle (EREV), also referred to as a Range Extended Electric Vehicle (REEV) or sometimes as a Range Extender (REX), is considered a "series hybrid." In a series hybrid configuration, the electric motor(s) is(are) generally operated instead of the engine and the motors are not "in-line" with the engine, so you can use one or the other and sometimes both at the same time. In this configuration, the electric motors always power the vehicles propulsion; however, the power source switches between the electric battery and the ICE with the engine getting engaged and operating like a generator once the battery has reached a predefined minimum state of charge and keeping the battery voltage maintained at that preset level until the vehicle is recharged from the grid (Figure 2).

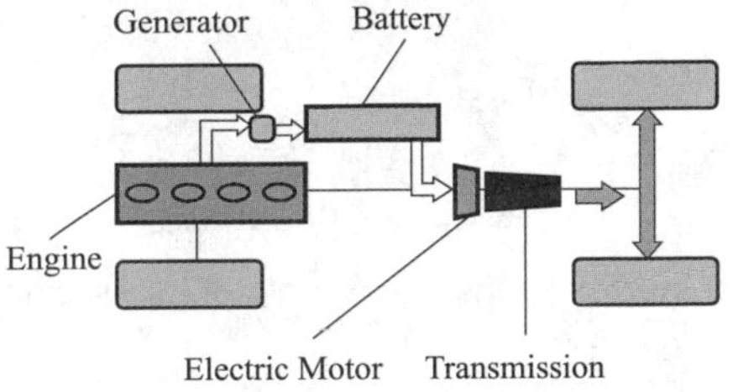

Figure 2 Series hybrid configuration

A common misconception of the PHEV and series hybrid configuration is that the engine will recharge the battery in this mode. This is not actually how they operate in the current designs. The engine will operate as a generator to power the electric motors but will only maintain the battery at its minimum state of charge until the vehicle can be plugged into the grid to recharge.

The term HEV refers to the most common configuration of electrified vehicles. These can range from the mild hybrid to a strong hybrid. A mild hybrid will have a smaller battery, usually less than one kilowatt hour (kW·h), and will provide less electric power to the system, whereas a strong hybrid will have a slightly larger battery, often about 1.5 kW·h, and will provide some minimal amount of electric propulsion in addition to powering some of the auxiliary systems. The term hybrid comes from the act of hybridizing of an ICE with a battery electric powered system, essentially creating a dual power system. In the hybrid car, the electric motor will generally provide power support throughout the operating cycle, but does not provide electric driving capability (mild hybrid) or at best only minimum electric drive capability (strong hybrid). During acceleration (Figure 3), the battery power will be added to that of the engine power to reduce the overall load on the engine, thereby improving fuel economy and reducing emissions. During deceleration, the motor recharges, or regenerates energy, back into the battery, which then provides power to the vehicle during start and stop events when the engine is turned off. It is important to understand that in the hybrid system the electric power to charge the battery is entirely provided by the on-board generator, the HEV does not plug into the grid to charge the battery.

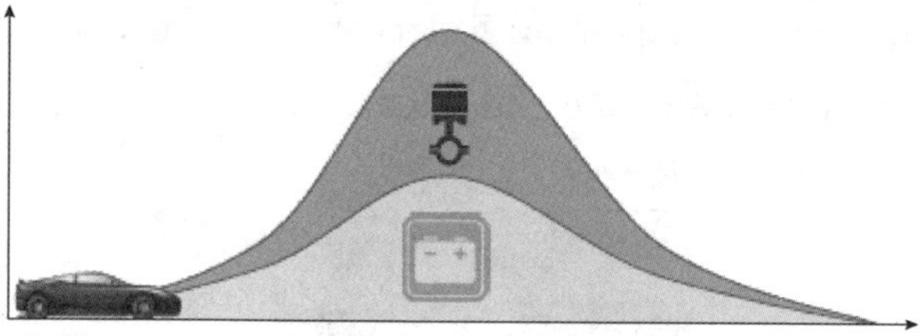

Figure 3　Operating cycle of hybrid car

A microhybrid (μ-HEV) refers to a Stop/Start (S/S)-type system. In this type of hybrid, there is a very small battery, often a lead acid or a very small lithium-ion battery, that powers the vehicle systems when the control system automatically shuts off and then restarts the engine when the vehicle stops. Most of these systems do not offer support for regenerative braking because the battery is too small to capture the amount of energy that would be regenerated back into it. However, some systems with larger batteries (usually those with >750~1000W·h) may include regenerative braking capability and on-board power support for electrically powered systems such as power steering, power brakes, entertainment system, HVAC, and vehicle lighting while the engine is stopped. This configuration is gaining much popularity in all regions around the world due to the low cost and general simplicity of this system.

A Neighborhood Electric Vehicle (NEV) is a fully electric vehicle but it is limited to a maximum speed of about 30mph (up to 45mph in some areas), must have a gross vehicle weight of less than 3000pounds, and can be driven only on limited public streets. These limits are based on current US laws and regulations for this type of low-speed vehicle. The NEV is commonly used on college campuses and in small private communities. These vehicles can basically be considered next generation or "souped up" golf carts as that is their basic beginning structure. One important thing to take note of is that this same term has a very different meaning in different regions of the world. In China, this same term, NEV, is used to describe China's New Energy Vehicle (NEV) policy. This policy was developed in China based on the "863" program and subsequently the Tenth Five Year Plan, which created a set of incentives and goals for increasing the amount of New Technology Vehicles in an effort to reduce air pollution. This policy was also developed in order to advance Research and Development (R&D) in the areas of HEVs, EVs, and Fuel Cell Vehicles (FCVs) in China. For the purposes of this text the term NEV will be used to represent the Neighborhood Electric Vehicle unless otherwise stated.

There is another term that is occasionally used in reference to electric vehicles, that is Light Electric Vehicle (LEV). The LEV ranges from both two- and three-wheeled electric motorcycle- type platforms to electric bicycles. The LEV market is actually the largest electric vehicle market in the world with between 25 and 30 million LEVs sold annually throughout the world. The vast majority of these are electric bicycles (e-bikes), pedelecs (electric assist bicycles), and electric scooters that are sold in Asia, but demand is growing strongly in the Western markets for this type of technology.

There are also a couple of EV industry terms that should be understood. The California Air Resource Board (CARB) or sometimes referred to as the Air Resource Board (ARB) is one of the biggest reasons for the growing demand for electric vehicles in the United States. In 1990, CARB passed into legislation the Zero Emissions Vehicle (ZEV) mandate, which required that by 1998 two percent of all vehicles sold in California must emit zero emissions, increasing to 5% in 2001 and 10% in 2003 (California Air Resource Board, 2014). While this early mandate for vehicle electrification missed its initial targets, it has become the driving force for both electrification and hybridization in the United States. By 2009, there were at least 21 other states that had either adopted ZEV mandates or were considering doing so. In 2013, eight US states including California, Connecticut, Maryland, Massachusetts, New York, Oregon, Rhode Island, and Vermont signed a memorandum of understanding (MOU) to put 3.3 million ZEVs on the roads by 2025 through a combination of R&D, infrastructure, education, and incentivization. This move would increase ZEV penetration to about 15% of all vehicle sales (Figure 4) in the United States.

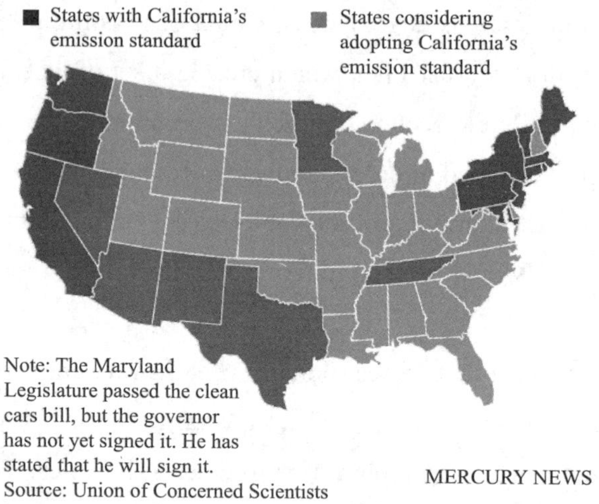

Figure 4　States that have enacted ZEV emissions standards

The other legislative reason automakers began looking at a portfolio of solutions to improve fuel economy was the US Corporate Average Fuel Economy (CAFE) standards that were enacted in 1975. These standards were originally put in place as a result of the Organization of the Petroleum Exporting Countries (OPEC) oil embargo of the early 1970s in an effort to reduce the US reliance on foreign oil. Initially the law required that automakers passenger fleets achieve 18 miles per gallon (MPG) by 1978, increasing to 27.5mpg by 1985. However, during the Regan administration, the law was relaxed down to 26mpg and then increased again to 27.5mpg by 1990 and then for over 21 years the CAFE laws remained fixed at 27.5mpg. In 2006, the CAFE laws were reformed and a new manner of measuring CAFE was installed in 2011, by vehicle footprint instead of weight. And a new set of fuel economy targets were put in place between 2011 and 2015 raising the target from 27.5mpg up to 35.5mpg by 2015 (2016 model year). In 2012, the next generation of standards were agreed to that define the required improvements between the 2017 model year and 2025 model year, resulting in a fleet average fuel economy in 2025 of 54.5mpg.

2 Stationary and Grid Terminology

We will also briefly describe several of the key terms that are used in the large stationary energy storage market before we delve into the battery specific terminology. Stationary and grid-based energy storage generally refer to the large, often megawatt-sized systems that utility companies put in place in order to provide several functions to support the grid operation. These functions include providing storage for intermittent and renewable generation sources such as wind and solar (renewables), providing fast demand response, load leveling, managing the power quality, and ensuring reliability of their power generation services.

The term Community Energy Storage (CES) refers to a relatively small energy storage device that often ranges from about 25kW·h up to about 100kW·h in size. This technology is intended to be used by utility companies to install capacity in specific neighborhoods where they see increased demand and/or that are at the periphery of the grid. This enables both "islanding." the ability to operate a microgrid independent of the larger grid, as well as support during downed power events. One organization, the Energy Storage Association (ESA), has published an "open-source" set of requirements for these types of application in an effort to begin industry standardization and communization (Energy Storage Association, 2014).

The CES falls into the category of energy storage systems (ESSs) referred to as Distributed Energy Storage (DES) systems. In general, the DES refers to systems much larger than the CES ranging from hundreds of kilowatt hours up to megawatt hours (MW·h) of installed capacity. The DES systems are generally installed at utility-owned location such as the electrical substations and offer some of the same benefits as the CES systems, only at a much greater scale.

3 Battery Terms

Before we begin talking about battery design, we need to first understand some of the basic terminology surrounding the battery. Please note that this list is not intended to be an exhaustive list but rather is an explanation of some of the most common terms that you will find in the lithium-ion battery industry. For a more exhaustive list of terms and definitions, again I would refer the reader to one of the industry standards organizations such as the SAE standard J1715 “Battery Terminology”.

Ampere—Often referred to as “Amp.” this is a unit of measurement of the battery current.

Anode—The anode is the negative “–” terminal inside the battery cell. It is typically a thin piece of highly conductive aluminum or copper that is coated with graphite, carbon, or other similarly conductive material.

Battery Management System (BMS)—The BMS is the control system within the battery pack that consists of one or more electronic controllers, which manages the charging and discharging, monitors the temperature and voltage, communicates with the vehicle system, balances the cells, and manages the safety functionality of the battery pack.

Beginning of Life (BOL)—The term BOL refers to the battery energy, capacity, and power when it is first built or is at the beginning of its life.

C-rate—The term C-rate is important; it refers to the rate at which a battery can charge or discharge all of its energy (or power). In other words, it describes how fast a battery can accept a charge or give up its power (discharge). C-rate is described in relation to a 1h discharge, so 1C-rate is equal to the rate at which a battery is fully discharged (or charged) in 1h. Along the same lines, 2C-rate would then be equal to the rate at which a battery is fully discharged in 30min (60min/2C=30min). If the C-rate number goes up the discharge time goes down, and vice versa. So 0.5C discharge rate would be equal to a 2 h discharge period (60min/0.5C=120min or 2h).

Capacity—Capacity is measured in Ampere hours (Ah) and is a measure of amount of energy in a system. Think of capacity as being analogous to the size of the hose that water may flow through. Larger capacity is equal to a larger hose.

CARB—The CARB is a legislative group in California with the mandate of passing legislation that will improve the air quality of the state. CARB is often also referred to “ARB” for the Air Resource Board.

Cathode—The cathode is the opposite of the anode; it is the positive “+” terminal inside the battery cell. It is typically a thin piece of aluminum or copper that is coated with lithium-ion chemistry such as lithium-iron phosphate (LFP), lithium cobalt oxide

(LCO), lithium-nickel/manganese/cobalt (NMC), lithium-manganese oxide (LMO), or other lithium-based chemistry.

Current—Current is the measurement of the flow of electrical charge, which may be carried by electrons moving through a wire or circuit board and by ions moving through an electrolyte between the anode and cathode and is measured in Amperes.

Cycle—The term cycle refers to the process of discharging and then charging a battery. A complete discharge and then charge is known as one cycle. A cycle may run at various levels of power and/or voltage or even using a constant rate of charge and discharge, depending on the application requirements. A cycle may be full, completely discharging and then charging the cell, or partial, only discharging to a set level and then recharging back up to starting level.

Depth of Discharge (DOD)—The DOD is a measurement of how much of the cell or pack energy will be used for that application. You will typically use lithium-ion battery only somewhere between 20% and 90% of the total amount of energy in order to prevent overcharge at the top and manage low end voltage. So if we think of the battery as a 10-gallon gas tank PHEV, it may only use 60%, or 6 gallons, of its energy—so the DOD would be 80%. For a HEV or micro-HEV, the battery may only use 30%~50% of the available energy or 3~5 gallons in this example. And an EV may use 80%~90% of the total energy available, about 8~9 gallons in this example. During system development one design decision that is commonly made, especially for hybrid applications, is to reduce the Depth of Discharge (DOD) during operation in order to achieve greater cycle life and improve safety in the energy storage system.

Electrodes—The term electrodes refer to the combined pair of anode and cathode when it is assembled in a battery cell. Electrolyte—The electrolyte is liquid, gel, or other material that is used as the medium to transfer the lithium-ions back and forth between the anode and cathode.

Electric Miles per Gallon (eMPG)—eMPG refers the "electric miles per gallon" equivalent rating that is used by the US Environmental Protection Agency (EPA) to offer an equivalent comparison for consumers to traditional mpg used on ICE vehicles.

End of Life (EOL)—The EOL of a battery is reached when the battery's maximum power and energy have been reduced to about 80% of their BOL measurements.

The general rule of 80% is based on either power or energy that has dropped to a point where acceleration (power) or range (energy) is no longer “satisfactory” to the consumer. However, depending on the application, a lower EOL could be used.

Energy—The term energy, which is measured in kilowatt hours (kW·h), refers to the amount of energy that a battery will store, think of it as being analogous to the size of the gas tank.

Energy Density—Energy density is the measurement of how much energy a cell or pack contains in relation to either its mass or its volume. Energy density is measured either in watt hour per kilogram (W·h/kg) or as Watt hour per Liter (W·h/L). When it is used to compare to mass in the W·h/kg, it is referred to as Gravimetric Energy Density. When it is referred to in relation to volume in the W·h/L form, it is referred to as Volumetric Energy Density.

Energy Storage System—The term ESS is used in various forms, but it generally means the complete battery pack system. The ESS is the combination of cells that are mechanically and electrically connected, along with the appropriate thermal, electronics, and mechanical structure to house the entire unit. In essence, it is everything that is “in the battery box.”

High Voltage (HV)—any system with a voltage over 60V is considered “high voltage” and must include appropriate protections (HVIL, safety disconnect, orange cabling, etc.) in order to prevent harm to the workers or anyone who may come in contact with the system.

Impedance—Impedance is basically a measurement of how much the materials within the cell slow down, or impede, the flow of current when a voltage is applied. It is measured in Ohms and is represented by the symbol Ω. AC impedance is generally used in relation to measuring at the cell level. Impedance basically builds on the concept of DC resistance, which is a measure of magnitude (where there is no difference in phase shift between voltage and current), by adding the concept of phase to the measurement.

Jelly Roll—The jelly roll is the combined assembly of anode, separator, and cathode that are either stacked or rolled together and inserted into the can or pouch (Figure 5).

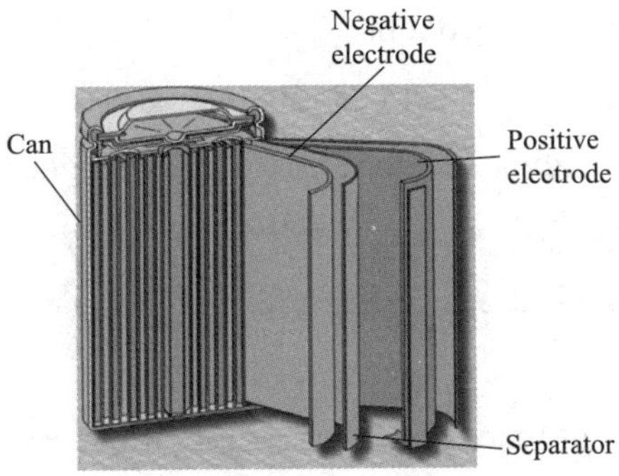

Figure 5 Schematic of a cylindrical lithium-ion battery

LiB—An acronym that is occasionally used in place of the terminology lithium-ion battery or li-ion.

Miles per Gallon—This is the standard measurement of fuel efficiency that is applied to almost all ICE vehicles in the United States. In other regions, kilometers per gram CO_2 (kg/CO_2) is used for the same purpose.

Parallel—A battery with parallel connections refers to cells that are connected in parallel (e.g., positive to positive, negative to negative, etc.). In a parallel connection, you are feeding current into all of the cells at the same time and pulling current out of them at the same time. When connecting cells in parallel, the system capacity is increased. The example below represents three cells: let us assume that they are 3.6V and 5A·h; in a parallel configuration, this would end up with still at 3.6V but the capacity would increase to 15A·h (5A·h×3cells) (Figure 6).

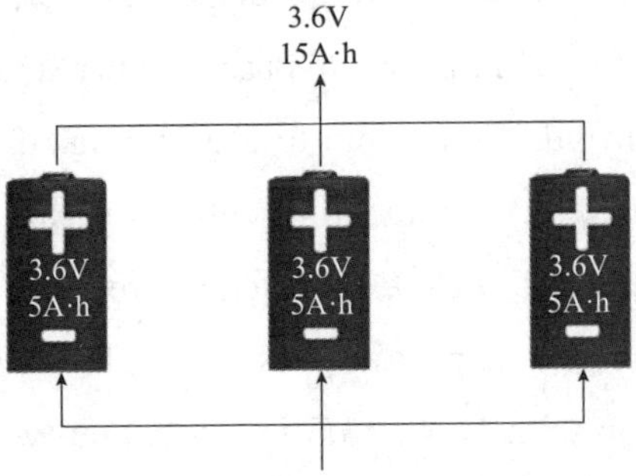

Figure 6 Parallel cells

Power Density—Measured in either kilowatt per kilogram (kW/kg) or kilowatt per Liter (kW/L). Similar to energy density, power density is the comparison of a battery's power in relation to its weight or volume.

Power Net—The term power net is a relatively new one. It is getting used largely in European countries by automakers to refer to the collection of on-board technologies and software that is being run electronically. It usually includes the infotainment systems, radio and communications systems, navigation systems, and anything else that is electrically powered.

Primary Battery—A primary battery is simply a battery that is not rechargeable. An example of a primary battery would be the AA-, C-, or D-type battery, which you might use in your household electronics that are often made using an alkaline chemistry.

Resistance—Resistance is related to impedance and is a measurement of how much the materials within the cell slow down, or resist, the flow of current when a voltage is applied. It is measured in Ohms and is represented by the symbol Ω. DC resistance is generally used in relation to measuring at the pack level. DC resistance describes the magnitude of the difference between voltage and current (assumes no difference in phase shift between voltage and current).

Secondary Battery—A secondary battery is simply a battery that is rechargeable; examples include lithium-ion, nickel-metal hydride, and lead acid.

Separator—The separator is a thin sheet of material, often a single or multilayer plastic (polypropylene) or ceramic based material, that separates the anode from the cathode in order to prevent them from touching and creating a short circuit. The separator also must allow the lithium-ions to pass between anode and cathode.

Series—A series configuration refers to a grouping of cells connected in series (e.g., negative to positive, etc.). Connecting cells in series increases the voltage of the overall system. The example below represents three cells: let us assume that they are 3.6V and 5Ah similar to our previous example, in a series configuration. This configuration would end up with 10.8V (3.6V×3 cells) but will remain at 5Ah. Putting lithium-ion cells in series is similar to connecting multiple garden hoses end to end, just as the garden hoses are in series so are the batteries. The difference is that in batteries connecting them in series increases the voltage but not the capacity—the same effect does not occur when connecting garden hoses in series (Figure 7).

Figure 7 Lithium-ion cell in series connection

Short Circuit—A short circuit occurs when the positive and the negative poles or electrodes of a battery are connected. In essence, a short circuit creates a circular connection within a cell, driving all of the current back into the cell or pack which will eventually, and usually very quickly, lead to a catastrophic failure. This could occur inside the cell due to the growth of dendritic materials that grow between the anode and cathode connecting them electrically. This could also happen if a microscopic piece of debris gets assembled in the jelly roll, which could eventually pierce the separator and connect the two electrodes. If it happens inside the cell, it is an "internal" short; but if the electrical connection is made between the poles outside of the cell, it is referred to as an "external" short (e.g., outside of the cell).

State of Charge (SOC)—While DOD measures how much of the battery is being used, SOC measures how much is left at a specific point in time. SOC is usually measured from 0% to 100%, but it is measuring against the DOD. Basically the SOC tells you how much energy or power you have right now, it is like the gas gauge in your car (Figure 8).

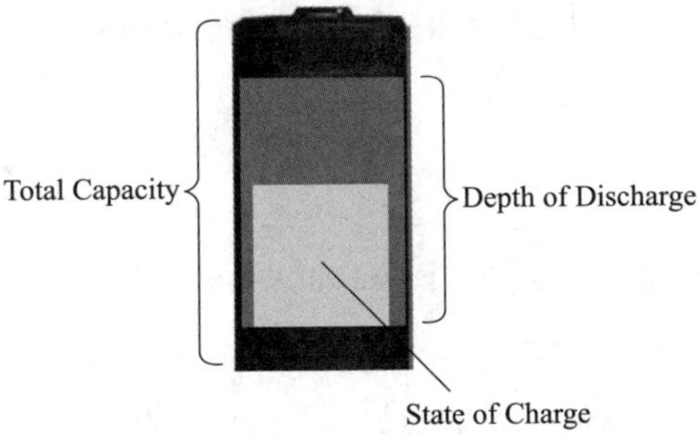

Figure 8 DOD, SOC, and total capacity of a lithium-ion cell

State of Health (SOH)—SOH is an interesting measure because it does not represent any "standard" measurement; different battery companies, controls companies, and application users may have different definitions. In general, SOH refers to the current state of health of the battery as compared to its beginning of life measurement. In other words, SOH is intended to tell you how long the battery will take to reach its end of life (EOL). In essence, SOH is a measure of internal resistance, capacity, voltage, self-discharge, the battery's ability to accept charge, and the total number of charge-discharge cycles that the battery has completed at that point in time. The state of health calculation is an algorithm that is programmed into the battery management system main controller.

State of Life (SOL)—The SOL is often used synonymously with the state of health. However, it may also be a separate algorithm that is intended to determine how much life is remaining in the battery based on some of the same measurements as are used in the state of health calculation.

SO_x—The term SO_x is used to represent all of the "State of" calculations at the same time, including state of health, state of charge, state of life, etc.

Voltage (V)—Voltage is the potential of the charge in a battery, but for clarity it can be thought of as being analogous to the pressure in a hose.

Selected from Chapter 3 of "*The Handbook of Lithium-Ion Battery Pack Design: Chemistry, Components, Types and Terminology*", John Warner, Elsevier, 2015.

Thermal Management

1 Introduction

Thermal management of a battery is, in its most simple form, the integration of some form of heat exchange device into the battery system in order to maintain the battery cells at a near constant temperature. There are many potential methods that can be used to achieve this, but it typically involves moving heat through some medium away from the cells and out of the pack. The complexity of the thermal management system will be highly dependent on three factors. The first is the duty cycle under which the pack will operate. If it is intended to be used as a high-power application, the battery cells will generate more heat than if it is used as an energy application. The second factor is the region where it will be used in. If the pack is intended to be used in a region of the world that frequently has very high ambient temperature, it means that the batteries will begin their cycle at an already elevated temperature level. Finally, the third factor is the cell itself. Different chemistries perform differently under high load and temperature operation. Therefore, it is important to consider all of these factors when beginning to design your battery's thermal management system.

The thermal management system should be able to maintain a temperature difference of about 2~3°C from the coolest cell to the warmest cell. At the worst-case condition, usually for larger packs the difference can be as much as 6~8°C. The reason why this is important is that a large temperature gradient between the cells will cause the cells to age at different rates. So the hotter cells will age faster than the cooler cells, and if there is a large gradient, this could mean that the battery's calendar life will be reduced prematurely.

There are three different types of heat transfer that need to be considered in battery design: conduction, convection, and radiation. Conduction refers to a direct transfer of heat energy from two objects that are in direct contact. Convection occurs when heat is conducted through a liquid medium to a heat-sinking device. Radiative heat transfer refers to heat energy that generated through electromagnetically thermally charged particles of matter that radiate from one source to another, generally through the air. All three methods of heat transfer must be considered in the battery system design, but conduction and convection will have the greatest impact on the thermal system design.

For example, as the cells are discharged they generate heat, that heat will be transferred via conduction to the bus bars and any other components that are in direct contact with the cells. Convective heating and cooling are most typically seen through the use of a liquid cooling plate; however, the movement of cooled air can also provide convective cooling. This is where most of the cooling will take place in the battery system. Finally, radiant heating must be considered in a couple of important areas. First, the radiated heat from the cell to other components that are not in direct contact with the cells can impact adjacent cells. Second, the heat that is being generated from components such as the electronics will impact the cells if it is not properly managed.

Sources of heat generation inside a battery system come from the chemical reaction within the cell (the main area of heat generation) as the lithium-ions move back and forth during operation; balancing of cells (usually while parked); electronics within the battery pack; and thermal management system. While the lithium-ion cells are the primary source of heat generation inside a pack, the influence of the electronics must not be disregarded as it can be a significant source of heat creation that must be accounted for in your thermal management system design. If adequate shielding and consideration of the placement of these electronics are not considered, then the heat generated by the electronics may actually have a negative impact on the cells life.

Figure 1 offers some simple examples of each of these three methods of heat transfer. In this example, a pouch-type lithium-ion cell is shown with a cooling plate on the right and the adjacent mounting frame on the left. The heat transfer from the cell to the cooling plate uses conduction both in the direct conduction of heat from the cell to the cooling plate and to the liquid cooling channels. These channels then use convection to remove the heat through a liquid medium away from the cells. At the top of the cell, a simple copper bus bar is shown being connected to the tab of the cell. This demonstrates the heat transfer through direct conduction from the cell to the bus bar. Finally, the frame piece to the left of the cell may be used for mechanical structure of the module; however, even if it is not directly in contact with the cell it will experience radiative heating from the cell.

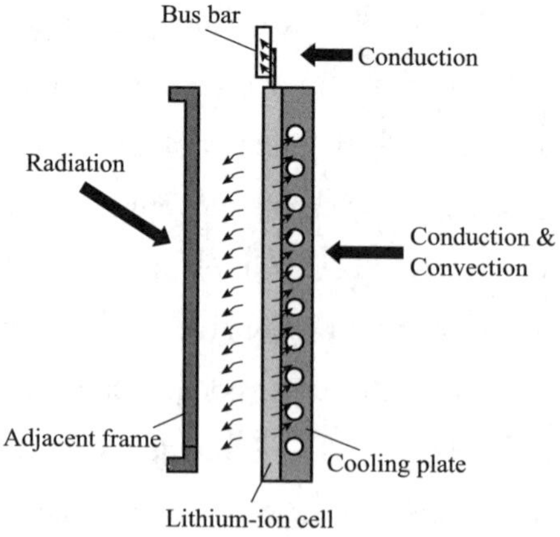

Figure 1 Sources of heat in a lithium–ion battery

Lithium-ion chemistries tend to operate best between about 10 and 35°C; this is referred to as the optimal temperature range. This is where you want the batteries to be at most of the time. However, most all lithium-ion chemistries will still operate down to about −20°C and up to about 45°C; this is known as the operational range. In this temperature range, no reduction in battery life would be expected to be experienced during normal operation. Between −20 and −40°C the electrolytes may begin to freeze and the cold temperatures increase the impedance within the cell thereby resisting the flow of ions and reducing capacity and performance, and above 60°C many lithium-ion cell chemistries begin to get more unstable; this is known as the survival temperature range (Figure 2).

This brings us to another concept that we should consider, at least briefly, in our thermal system design—thermal mass. Thermal mass is essentially the amount of heat that each object within the pack that has mass (all of them!) can absorb. Much in the same way that a battery can only store so much energy, all of the components in the battery pack can absorb a certain amount of heat, which is directly relative to their mass and the specific thermal properties of that particular material. For example, a copper bus bar will be able to absorb more heat than a plastic cover. Above its maximum ability to absorb heat, the material will begin to breakdown, plastics will melt and/or burn, and metals will begin to melt. Thermal mass will not generally drive major design considerations, other than in some of the components that are directly in contact with the areas of heat generation. However, in a passively cooled pack or an uncooled pack, this concept must be examined to evaluate its impact. Heat capacity may also come into effect if the application is being used in regions of high ambient temperatures. This is because in these regions the battery pack may absorb some of this ambient heat, which will cause the thermal system to have to work much harder in order to remove this heat. An example of this may be seen in regions such as the state of Arizona in the United States. Here, with summer ambient temperatures that may be in range of 38~44°C (100~110°F) or higher, the battery pack will absorb that heat and therefore will have to work to cool down the pack from these initial high temperatures back down to 23°C in order to preserve the life and performance of the battery system.

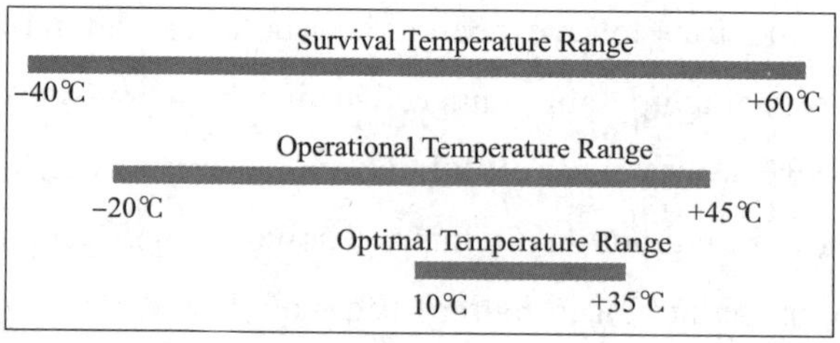

Figure 2 Lithium-ion cell temperature ranges

At temperatures above about 90°C, a polymer-based separator may begin to melt and breakdown, and between 90 and 130°C the separator will continue to breakdown until a series of internal short circuits between the anode and the cathode are experienced; at this point, the cell will begin to move toward what is known as “thermal runaway”. In effect, thermal runaway means that the cell becomes hot enough to create self-sustaining heat generation and failure is imminent in the form of cells venting and/or explosion (often referred to as “rapid disassembly”). There is no way to stop a cell in thermal runaway once the threshold has been surpassed. The exact temperature that a cell reaches the thermal runaway threshold is different for different chemistries. Some may reach it as low as 120°C, while others may be able to exceed 140°C before reaching this event.

The other factor that occurs during a thermal runaway is that as the chemistry inside the cells begins to breakdown, side reactions begin occurring rapidly generating oxygen. This oxygen generation may continue to fuel the burning of the cell. In this instance, a sealed pack design may offer some benefit due to the fact that no new oxygen is allowed to enter the pack. Therefore, once the oxygen inside the pack is consumed, there may not be enough oxygen to sustain a flame. One final comment on thermal runaway is that the temperatures that can be experienced during these events may be able to exceed 600~800°C depending on the size of the cell and how many cell fail. There are few solutions that can prevent a cascading failure in the event of a thermal runaway with current chemistries. However, system design should take these safety factors in account as much as possible and attempt to either manage to thermally isolate the failing cell and/or manage the venting of the cell. In any case managing cascading cell failures needs to be looked at as a complete system solution. There is no "silver" bullet that can by itself prevent cascading cell failures.

There are a couple other terms that we should briefly discuss in relation to thermal management systems design and testing, since they require a bit of explaining to clearly understand. The first one is the term adiabatic. The adiabatic process is related to the first law of thermodynamics that basically means that there is no transfer of heat from the battery to the outside environment. In relation to lithium-ion batteries, this may be used in relation to thermal characterization testing on a lithium-ion cell. For example, when the cell is being tested, it is done in a thermal chamber that has the capability to operate at varying temperatures. So when the cell is being tested, the thermal chamber is maintained at the exact same temperature as the cell throughout the test. This allows for the accurate definition of how much heat is generated by a cell during a specific operational cycle.

The other two terms that I want to explain are exothermic reactions and endothermic reactions. An exothermic reaction is one that occurs when energy is discharged and heat is released into the environment thereby causing heating. An endothermic reaction occurs when energy is absorbed and heat is absorbed from the environment thereby causing cooling. In lithium-ion cells, exothermic reactions occur when the cell is being discharged (releasing energy) thus generating heat that must be managed through the thermal management system of the battery.

2 Why Cooling?

As mentioned above, in order to maximize the potential of lithium-ion cells, they need to be maintained at about 23~25°C (73~77°F) throughout the majority of their use cycle. However, under operation, the cells experience an exothermic reaction—they begin generating heat due to the rate at which the chemical reaction occurs within the cell and the related increase in cell resistance. This reaction in combination with high ambient (outside) temperatures means that your battery design must be able to cool the batteries down and maintain them within their optimal operating range in order to ensure the performance and life of the overall battery system.

Additionally, high discharge rates generate the exothermic temperature increases within the cells. And when these discharges come frequently, it means that the cells do not have time to cool down between these pulses, which again drive higher temperatures. In addition to this, there is less time for the thermal management system to engage and reduce the temperature of the cells back down. Think of this as a stair-step effect: when you hit your accelerator, a rapid discharge of the battery occurs; with frequent stopping and accelerating events the battery thermal management system will not have time to cool down the battery from the last discharge, causing the battery to gradually but steadily increase in temperature. An example of this is the traditional HEV, which will continually discharge and then regeneratively charge the battery regularly during a driving cycle. In this type of usage cycle, the battery may not have time to cool down the cells before another discharge-charge cycle is initiated. This will cause a slow and steady increase in the temperatures within the pack. Allowing the pack to sit unused for a period of time will generally allow the system to reduce the temperature back down to its normal operating range. Figure 3 below offers an example of what this type of temperature increase may look like. While this is not actual performance data, it is consistent with the type of heat generation that could be expected to be seen in this type of application.

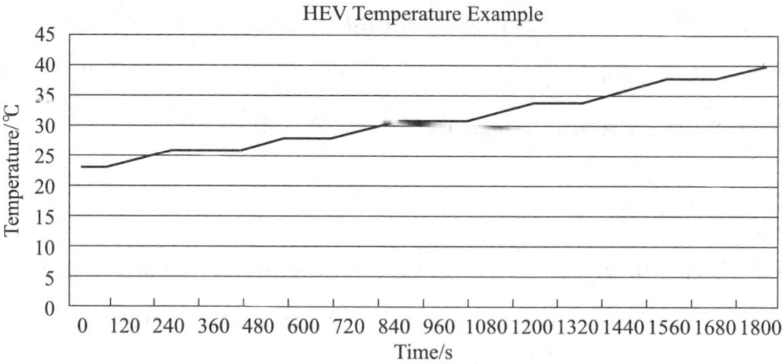

Figure 3 HEV temperature example

As mentioned earlier, in high-temperature environments where the ambient air temperature is already in the 30~35°C range or higher, it becomes even more important to be able to reduce the temperature down below the ambient in order to ensure life, performance, and against catastrophic failure. Constant operation at high temperatures reduces cell life and speeds the aging process. This was demonstrated in 2012 when a group of Nissan Leaf owners in the Arizona region complained to Nissan about the low electric range and reduced battery capacity that they were getting with their vehicles after only a short period of time. While there were not initially many of these complaints, there were enough to generate much unwanted national media attention for both Nissan and the industry in general.

In essence, Nissan's evaluation of the vehicles and the way they were being used found that the users who were experiencing this reduction in capacity tended to be driving about 150% of the average US annual mileage, some of whom were also using frequent fast charging.

So even while the batteries were performing to their design specifications, the continuous high temperatures and more extreme usage drove a reduction in battery capacity and therefore reduced range. And even though Nissan's owner's manual tried to clearly identify the major areas that will reduce battery life, they suffered much undeserved negative attention (Figure 4).

CAUTION

To prevent damage to the Li-ion battery:

- Do not expose a vehicle to ambient temperatures above 120°F (49℃) for over 24 hours.
- Do not store a vehicle in temperatures below −13°F(−25℃) for over seven days.
- Do not leave your vehicle for over 14 days where the Li-ion battery available charge gauge reaches a zero or near zero (state of charge).
- Do not use the Li-ion battery for any other purpose.

Figure 4 2012 Nissan LEAF Owner's Manual battery warning

In the end, Nissan ended up changing their warranty between the 2012 and the 2013 model year vehicles in order to account for these types of occurrences. Prior to the 2013 model-year vehicles, the warranty specifically excluded gradual loss of charge/capacity. To their original 8-year/100000-mile warranty that covers defects and flaws, they added a 5-year/60000-mile warranty that covers capacity. If the Leaf loses more than 30% of its original capacity in the first 5 years or 60000 miles, Nissan will repair or replace it.

In addition, Nissan began working on the release of a new battery pack design that was intended to resolve these issues. To resolve these issues, Nissan began looking at alternative battery chemistries that would perform better in the high temperature regions of the world. Additionally, in mid-2014, Nissan announced that they would be offering replacement battery packs for earlier models at a price of $5499 USD (O'Dell, 2014). This is most interesting as this calculates out to a price of about $229 per kW·h ($229/kW·h), which is much below the current market pricing. Nissan is very likely taking a loss on these replacement packs in order to help inspire current and future customers, but it will be at a cost.

To tie this back in with the thermal management system of the battery, the reader should recognize that the Nissan Leaf battery pack (as will be discussed in more detail in a following chapter) actually has no active thermal management system at all. Instead, the Leaf battery pack designed by AESC uses the metal enclosures of the module and the pack to act as a heat sink to spread the heat out. But in hot climates or in frequent rapid discharges, there is no means to increase the rate of cooling in order to more rapidly respond to the increasing temperatures. While this provided a low cost and fast to market solution, as we have seen it also caused some other problems on which they are still working hard to resolve.

Finally, we should talk about the high-voltage electronics and balancing systems when we talk about thermal management. In many systems, the high-voltage electronics can be a major portion of the heat generation of the pack, especially in smaller packs. Additionally, cell balancing by its very act is the transference of electrical energy into heat energy. In essence, as you balance the pack you also heat it up. In the cases where level 3 charging is used, most battery systems are designed to turn the thermal management system on during these charging events due to the rapid heat generation that will otherwise ensue. Without the cooling system being engaged during a fast-charge event, the battery may generate enough heat to reduce the amount of energy it will be able to hold and could generate enough heat to actually cause the battery system to shut down until it cools. Think of this as trying to use a fire-hose size nozzle to fill your gas tank; the tank will simply not be able to take that volume of fuel at once. In the battery system, it is the not the speed but in fact the heat that is being generated that causes the problems and stresses the lithium-ion cells.

In the case of the electronics within the battery system and their heat generation, it is generally easy enough to run cooled air across the electronics to help cool them down. However, the designer in these instances needs to ensure that the air is flowing to the lithium-ion cells either before or in parallel with the flow to the electronics. Most electronics have a much wider operating temperature range than the batteries do, so it is more efficient to cool the cells first and the electronics second. In liquid-cooled battery packs, it is somewhat more difficult to ensure the cooling of the electronics unless cooling plates are designed into the pack. But in both situations the location and placement of the electronics require special attention by the energy storage system designer. All of these need either some amount of thermal management or at least to have a system design that does not negatively impact the performance of the batteries.

3 Why Heating?

On the other end of the spectrum is heating the battery. As a general rule, battery system designers do not want to heat a battery except for at very low ambient temperatures and even then it should not be a "rapid" heating.

Most liquid electrolytes used in modern lithium-ion batteries will begin to experience reduced power at about −10°C and will begin to freeze at between −20 and −30°C, which makes them unable to provide power at very low temperatures.

In a liquid-cooled system, a heat pump can be added to the overall system to provide warmed liquid through the cooling loop, which will slowly heat up the batteries. Other methods may also be employed such as using a thin-film heater. In this instance, the heater may not actually be used to heat, but rather to slow the rate of cooling such that if the vehicle sits for a weekend the cells will only be down to 25°C after several days. This means that the system would have full performance, however, would suffer from some energy loss as the thin-film heater would need to be powered by the battery itself. However, this would be only a minor and temporary capacity loss.

4 Active Thermal Management Systems

Active thermal management involves using some medium such as air, liquid, or refrigerant that is forced through the pack and over the cells to reduce temperatures. The two most common methods are air cooling using chilled air that is directed throughout the pack and over the cells and electronics to reduce the temperature of the cells and liquid cooling. This generally requires the integration of a fan, ducting, and heat transfer plates of some sort. The benefits are that it can be relatively effective in responding to rapid changes in temperature and has a lower weight than a liquid-cooled system. The other benefit is that the cooled air is directly flowing across the cells. The disadvantages are that air is not as effective a cooling medium as liquid. And depending on air-flow design, it can cause the cells at the beginning of the air flow to be cooler than the cells at the end of the air flow because as the air passes by the initial cells, it begins to pick up the heat so by the time the air passes by the last cells it is warmer than when it in initially entered the pack. This uneven cooling can cause the cells to age at different rates, thereby reducing the life of the pack (Figure 5).

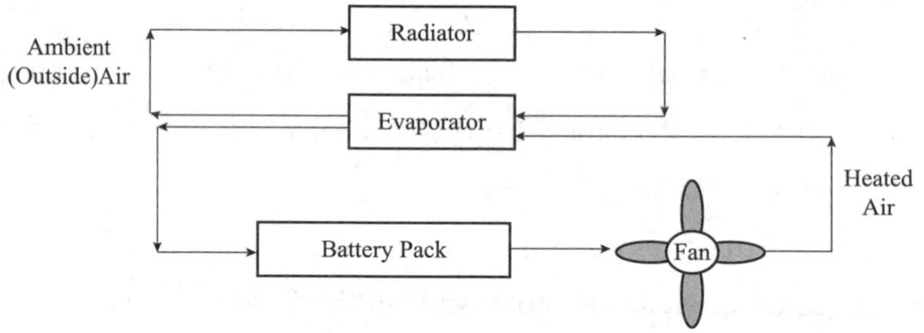

Figure 5　Active air cooling schematic

In addition to using chilled air, some systems will pull ambient air from the outside of the pack, or even inside the cabin, and circulate that air through the pack. In this case, the air must first be filtered and the cooling ability is limited to reducing the temperature only to what the ambient temperature is. So in this instance, if the ambient temperature is at 30°C the thermal management system will only be able to cool down the battery to 30°C (Figure 6).

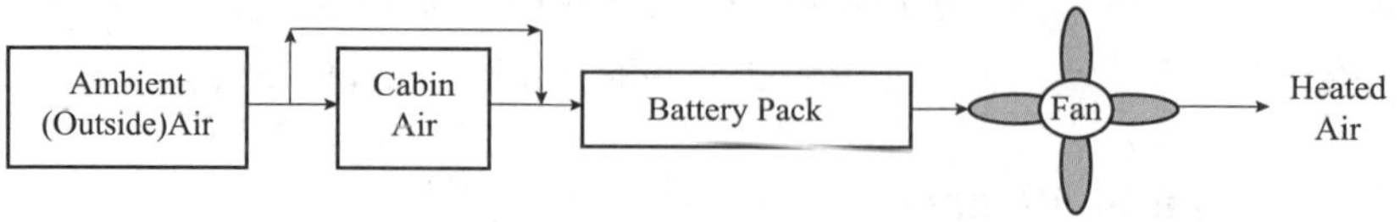

Figure 6　Passive air cooling schematic

Another challenge of air cooling is that it makes the system design more difficult in the event that the pack needs to be a "sealed" pack design. As the air-cooled pack is essentially an "open system", it becomes very difficult to achieve an IP69 sealing level on the pack, which makes it a very good solution for a pack that is mounted inside the vehicle or inside a building or container, but not as good a solution if it is mounted external to the vehicle.

The other common method of active thermal management is by forcing a liquid, often a 50/50 water and glycol mix similar to that used in engine cooling, through a series of plates that are mounted next to the cells. This system involves a liquid distribution system that must be integrated into the pack, most often done through a series of hoses and heat exchangers. The benefit of the liquid cooling system is that it is quite an effective medium for quickly transferring heat away from the cells. It can also, with a heating element in the system, be used to provide heated liquid in order to warm the battery in the cold weather. The disadvantages are that it tends to be a heavier system (greater mass) and there is always a risk of leaks in the battery pack. As the liquid-cooled system is essentially a "closed system", it tends to be much easier to seal to the environment making it a good solution for packs mounted external to the vehicle.

In the liquid-cooled pack, there are essentially two methods for managing the heat generation of the lithium-ion cells. In the first method, you can develop a plate that is affixed directly to the cells and flow the cooling/heating liquid directly through these plates (Figure 7).

The second method is to create a single plate through which the fluid flows, but rather than affix the cells directly to this plate, a series of "fins" are attached to the heat sync plate. The lithium-ion cells are then attached directly to these fins (Figure 8).

One other aspect to examine in actively cooling and heating lithium-ion batteries is direct and indirect cooling. These terms are sometimes used interchangeably with active and passive. Direct cooling refers to moving the cooling/heating medium directly across the cells, whereas indirect cooling/heating involves moving the cooling medium through a heat exchanger that is attached to the lithium-ion cells. In actual usage, direct cooling is much more feasible with an air-cooled system where the thermal system can be designed such that the air is either pushed or pulled directly across the cells.

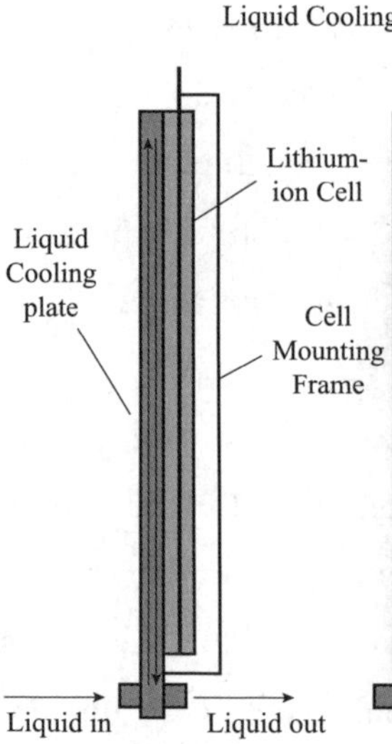

Figure 7　Liquid cooling plates

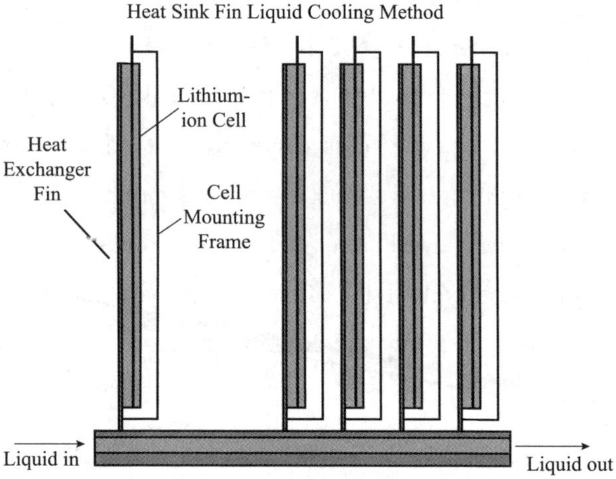

Figure 8　Heat sink fins

Refrigerant-based cooling systems are yet another method for thermally managing a battery system. Refrigerant-based systems have many of the same benefits as liquid-cooled systems, but tend to have a slightly higher cost. On the positive side, they eliminate the risk of liquid leakage into the battery pack. One such example is the system introduced by the Behr Group, now MAHLE International GmbH, who introduced a refrigerant-based cooling system targeted at HEV- and EV-type batteries (Figure 9 and Figure 10).

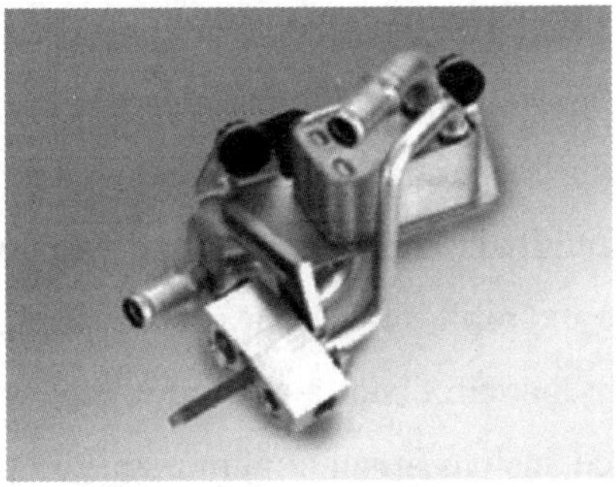

Figure 9 Behr refrigerant-based battery cooling system Courtesy MAHLE International GmbH

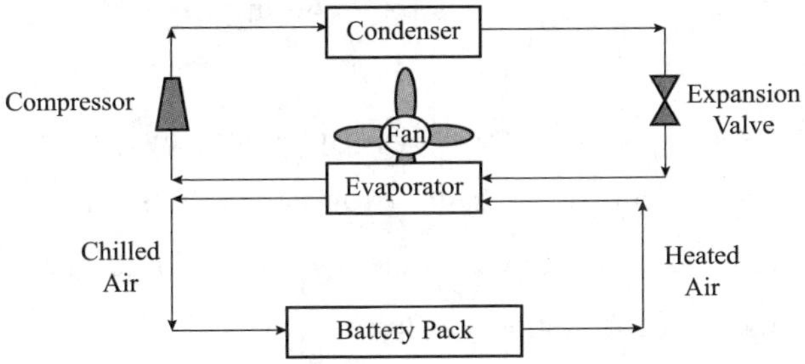

Figure 10 Direct refrigerant-based cooling schematic

Heat pipes have been discussed and evaluated for various lithium-ion-based battery thermal management systems, but as of today have not yet been able to offer either the performance or the cost advantage to make them worth investigating for large energy storage systems. In principle, the heat pipe combines two different thermodynamic principles, thermal conductivity and phase change, in order to transfer heat. The heat pipe operates by sealing a liquid into a series of metal pipes. As the pipes are heated, the liquid converts into steam and returns to the other end of the pipe where it again returns to a liquid form as it cools. Heat pipes have found significant market penetration in the portable computer market but have yet to find a home in automotive or large energy storage systems. The biggest downfall of the heat pipe for automotive applications is that above a certain temperature (depending on the liquid) all of the liquid is converted to steam and the system thereby stops transferring heat at all.

Another thermal management device that has been evaluated by some designers is the Peltier device. The Peltier device is a nonmechanical device that is made up of two different types of materials. When current is applied to one side of the device, heat is transferred from one side to the other where a heat sync can be used to move the heat out of the pack. This can be used to create what I refer to as a hybrid thermal management system. A hybrid thermal management system is one that may include various combinations of liquid, air, and other cooling mediums combined into a single battery pack system. One example may be a system design that seals the pack or module with an internal fan that circulates air within the pack; this would be combined with a heat exchange device such as an aluminum manifold or a Peltier device that the air is passed across. A cooled liquid that is entirely outside of the pack could then be run through the heat exchange device thereby cooling the air that is being circulated within the pack.

One more thing that must be considered in your thermal management system design is the pressure drop. The pressure drop is the difference between incoming and outgoing pressure of the fluid, which can be either liquid or air, and is basically a measure of the amount of pressure that the system must overcome in order to flow through the cooling system. In a liquid-cooled system, the pressure drop will determine the size and flow rate of the pump that is needed to move the fluid through the cooling system. In an air-cooled system, it will help to determine the size and flow rate of the fans that are needed to move the air through the system.

Selected from Chapter 10 of “*The Handbook of Lithium-Ion Battery Pack Design: Chemistry, Components, Types and Terminology*”, John Warner, Elsevier, 2015.

Appendix V. Nobel Prizes in Chemistry & More
（化学和相关学科的诺贝尔奖）

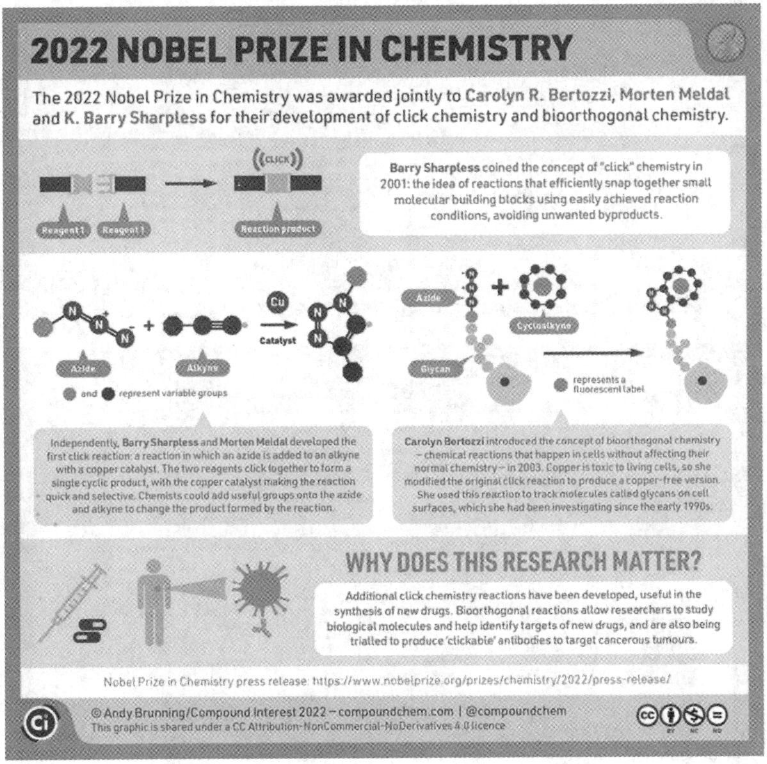

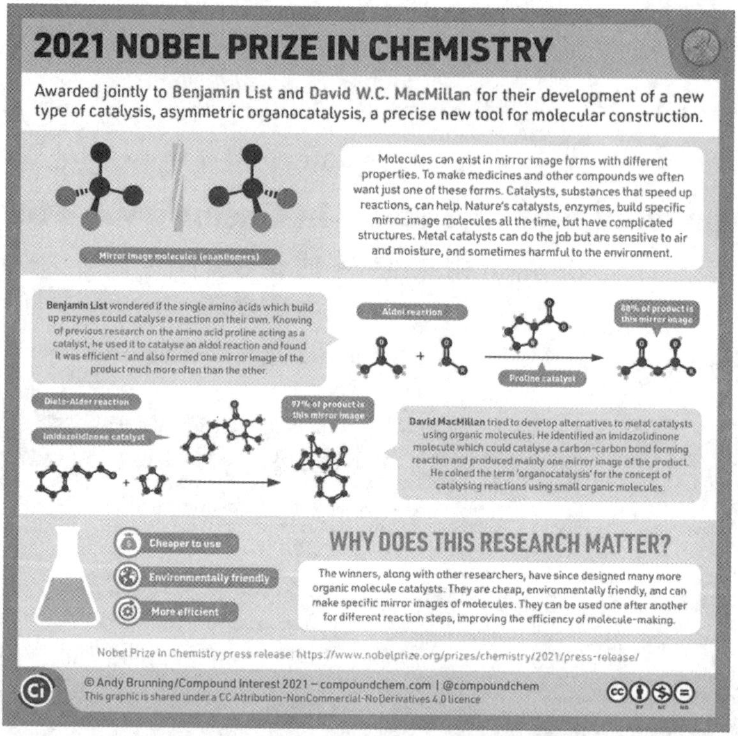
2021 NOBEL PRIZE IN CHEMISTRY
Awarded jointly to Benjamin List and David W.C. MacMillan for their development of a new type of catalysis, asymmetric organocatalysis, a precise new tool for molecular construction.
Mirror image molecules (enantiomers)
Molecules can exist in mirror image forms with different properties. To make medicines and other compounds we often want just one of these forms. Catalysts, substances that speed up reactions, can help. Nature's catalysts, enzymes, build specific mirror image molecules all the time, but have complicated structures. Metal catalysts can do the job but are sensitive to air and moisture, and sometimes harmful to the environment.
Benjamin List wondered if the single amino acids which build up enzymes could catalyse a reaction on their own. Knowing of previous research on the amino acid proline acting as a catalyst, he used it to catalyse an aldol reaction and found it was efficient – and also formed one mirror image of the product much more often than the other.
Aldol reaction
Proline catalyst
Diels-Alder reaction
Imidazolidinone catalyst
97% of product is this mirror image
David MacMillan tried to develop alternatives to metal catalysts using organic molecules. He identified an imidazolidinone molecule which could catalyse a carbon-carbon bond forming reaction and produced mainly one mirror image of the product. He coined the term 'organocatalysis' for the concept of catalysing reactions using small organic molecules.
Cheaper to use
Environmentally friendly
More efficient
WHY DOES THIS RESEARCH MATTER?
The winners, along with other researchers, have since designed many more organic molecule catalysts. They are cheap, environmentally friendly, and can make specific mirror images of molecules. They can be used one after another for different reaction steps, improving the efficiency of molecule-making.
Nobel Prize in Chemistry press release: https://www.nobelprize.org/prizes/chemistry/2021/press-release/
© Andy Brunning/Compound Interest 2021 – compoundchem.com | @compoundchem
This graphic is shared under a CC Attribution-NonCommercial-NoDerivatives 4.0 licence

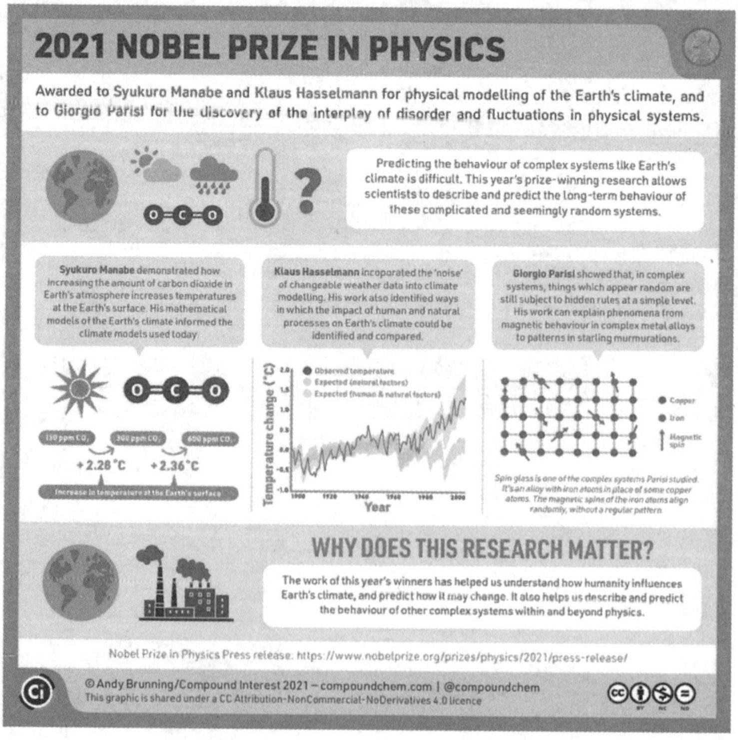
2021 NOBEL PRIZE IN PHYSICS
Awarded to Syukuro Manabe and Klaus Hasselmann for physical modelling of the Earth's climate, and to Giorgio Parisi for the discovery of the interplay of disorder and fluctuations in physical systems.
O=C=O
Predicting the behaviour of complex systems like Earth's climate is difficult. This year's prize-winning research allows scientists to describe and predict the long-term behaviour of these complicated and seemingly random systems.
Syukuro Manabe demonstrated how increasing the amount of carbon dioxide in Earth's atmosphere increases temperatures at the Earth's surface. His mathematical models of the Earth's climate informed the climate models used today.
Klaus Hasselmann incoporated the 'noise' of changeable weather data into climate modelling. His work also identified ways in which the impact of human and natural processes on Earth's climate could be identified and compared.
Giorgio Parisi showed that, in complex systems, things which appear random are still subject to hidden rules at a simple level. His work can explain phenomena from magnetic behaviour in complex metal alloys to patterns in starling murmurations.
+2.28°C
+2.36°C
Increase in temperature at the Earth's surface
Observed temperature
Expected (natural factors)
Expected (human & natural factors)
Temperature change (°C)
Year
Copper
Iron
Magnetic spin
Spin glass is one of the complex systems Parisi studied. It's an alloy with iron atoms in place of some copper atoms. The magnetic spins of the iron atoms align randomly, without a regular pattern.
WHY DOES THIS RESEARCH MATTER?
The work of this year's winners has helped us understand how humanity influences Earth's climate, and predict how it may change. It also helps us describe and predict the behaviour of other complex systems within and beyond physics.
Nobel Prize in Physics Press release: https://www.nobelprize.org/prizes/physics/2021/press-release/
© Andy Brunning/Compound Interest 2021 – compoundchem.com | @compoundchem
This graphic is shared under a CC Attribution-NonCommercial-NoDerivatives 4.0 licence

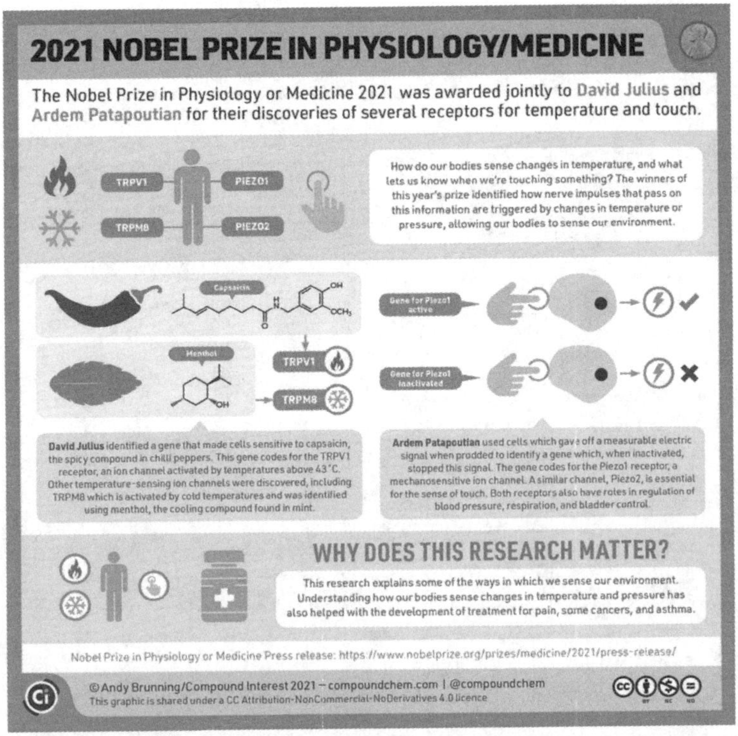
2021 NOBEL PRIZE IN PHYSIOLOGY/MEDICINE
The Nobel Prize in Physiology or Medicine 2021 was awarded jointly to David Julius and Ardem Patapoutian for their discoveries of several receptors for temperature and touch.
TRPV1
TRPM8
PIEZO1
PIEZO2
How do our bodies sense changes in temperature, and what lets us know when we're touching something? The winners of this year's prize identified how nerve impulses that pass on this information are triggered by changes in temperature or pressure, allowing our bodies to sense our environment.
Capsaicin
Menthol
TRPV1
TRPM8
Gene for Piezo1 active
Gene for Piezo1 inactivated
David Julius identified a gene that made cells sensitive to capsaicin, the spicy compound in chilli peppers. This gene codes for the TRPV1 receptor, an ion channel activated by temperatures above 43°C. Other temperature-sensing ion channels were discovered, including TRPM8 which is activated by cold temperatures and was identified using menthol, the cooling compound found in mint.
Ardem Patapoutian used cells which gave off a measurable electric signal when prodded to identify a gene which, when inactivated, stopped this signal. The gene codes for the Piezo1 receptor, a mechanosensitive ion channel. A similar channel, Piezo2, is essential for the sense of touch. Both receptors also have roles in regulation of blood pressure, respiration, and bladder control.
WHY DOES THIS RESEARCH MATTER?
This research explains some of the ways in which we sense our environment. Understanding how our bodies sense changes in temperature and pressure has also helped with the development of treatment for pain, some cancers, and asthma.
Nobel Prize in Physiology or Medicine Press release: https://www.nobelprize.org/prizes/medicine/2021/press-release/
©Andy Brunning/Compound Interest 2021 – compoundchem.com | @compoundchem
This graphic is shared under a CC Attribution-NonCommercial-NoDerivatives 4.0 licence

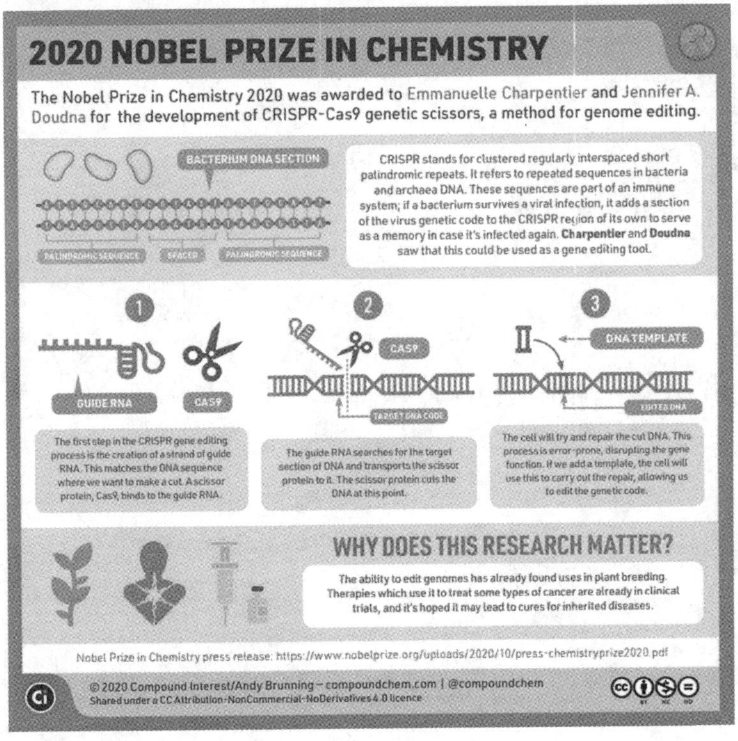
2020 NOBEL PRIZE IN CHEMISTRY
The Nobel Prize in Chemistry 2020 was awarded to Emmanuelle Charpentier and Jennifer A. Doudna for the development of CRISPR-Cas9 genetic scissors, a method for genome editing.
BACTERIUM DNA SECTION
PALINDROMIC SEQUENCE
SPACER
PALINDROMIC SEQUENCE
CRISPR stands for clustered regularly interspaced short palindromic repeats. It refers to repeated sequences in bacteria and archaea DNA. These sequences are part of an immune system; if a bacterium survives a viral infection, it adds a section of the virus genetic code to the CRISPR region of its own to serve as a memory in case it's infected again. Charpentier and Doudna saw that this could be used as a gene editing tool.
1
2
3
GUIDE RNA
CAS9
CAS9
TARGET DNA CODE
DNA TEMPLATE
EDITED DNA
The first step in the CRISPR gene editing process is the creation of a strand of guide RNA. This matches the DNA sequence where we want to make a cut. A scissor protein, Cas9, binds to the guide RNA.
The guide RNA searches for the target section of DNA and transports the scissor protein to it. The scissor protein cuts the DNA at this point.
The cell will try and repair the cut DNA. This process is error-prone, disrupting the gene function. If we add a template, the cell will use this to carry out the repair, allowing us to edit the genetic code.
WHY DOES THIS RESEARCH MATTER?
The ability to edit genomes has already found uses in plant breeding. Therapies which use it to treat some types of cancer are already in clinical trials, and it's hoped it may lead to cures for inherited diseases.
Nobel Prize in Chemistry press release: https://www.nobelprize.org/uploads/2020/10/press-chemistryprize2020.pdf
© 2020 Compound Interest/Andy Brunning – compoundchem.com | @compoundchem
Shared under a CC Attribution-NonCommercial-NoDerivatives 4.0 licence

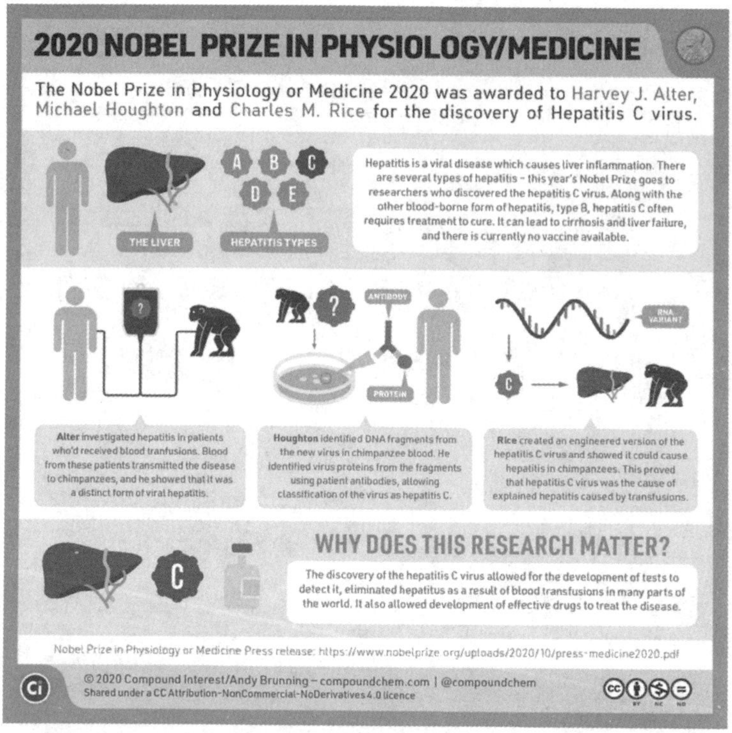
2020 NOBEL PRIZE IN PHYSIOLOGY/MEDICINE
The Nobel Prize in Physiology or Medicine 2020 was awarded to Harvey J. Alter, Michael Houghton and Charles M. Rice for the discovery of Hepatitis C virus.
THE LIVER
HEPATITIS TYPES
Hepatitis is a viral disease which causes liver inflammation. There are several types of hepatitis - this year's Nobel Prize goes to researchers who discovered the hepatitis C virus. Along with the other blood-borne form of hepatitis, type B, hepatitis C often requires treatment to cure. It can lead to cirrhosis and liver failure, and there is currently no vaccine available.
ANTIBODY
PROTEIN
RNA VARIANT
Alter investigated hepatitis in patients who'd received blood tranfusions. Blood from these patients transmitted the disease to chimpanzees, and he showed that it was a distinct form of viral hepatitis.
Houghton identified DNA fragments from the new virus in chimpanzee blood. He identified virus proteins from the fragments using patient antibodies, allowing classification of the virus as hepatitis C.
Rice created an engineered version of the hepatitis C virus and showed it could cause hepatitis in chimpanzees. This proved that hepatitis C virus was the cause of explained hepatitis caused by transfusions.
WHY DOES THIS RESEARCH MATTER?
The discovery of the hepatitis C virus allowed for the development of tests to detect it, eliminated hepatitus as a result of blood transfusions in many parts of the world. It also allowed development of effective drugs to treat the disease.
Nobel Prize in Physiology or Medicine Press release: https://www.nobelprize.org/uploads/2020/10/press-medicine2020.pdf
© 2020 Compound Interest/Andy Brunning - compoundchem.com | @compoundchem
Shared under a CC Attribution-NonCommercial-NoDerivatives 4.0 licence

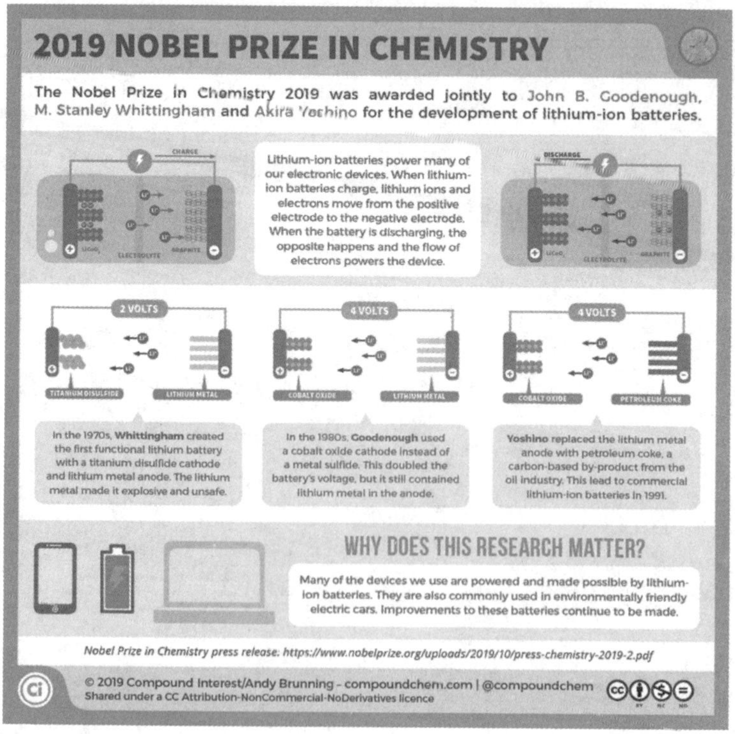
2019 NOBEL PRIZE IN CHEMISTRY
The Nobel Prize in Chemistry 2019 was awarded jointly to John B. Goodenough, M. Stanley Whittingham and Akira Yoshino for the development of lithium-ion batteries.
CHARGE
ELECTROLYTE
GRAPHITE
Lithium-ion batteries power many of our electronic devices. When lithium-ion batteries charge, lithium ions and electrons move from the positive electrode to the negative electrode. When the battery is discharging, the opposite happens and the flow of electrons powers the device.
DISCHARGE
ELECTROLYTE
GRAPHITE
2 VOLTS
TITANIUM DISULFIDE
LITHIUM METAL
4 VOLTS
COBALT OXIDE
LITHIUM METAL
4 VOLTS
COBALT OXIDE
PETROLEUM COKE
In the 1970s, Whittingham created the first functional lithium battery with a titanium disulfide cathode and lithium metal anode. The lithium metal made it explosive and unsafe.
In the 1980s, Goodenough used a cobalt oxide cathode instead of a metal sulfide. This doubled the battery's voltage, but it still contained lithium metal in the anode.
Yoshino replaced the lithium metal anode with petroleum coke, a carbon-based by-product from the oil industry. This lead to commercial lithium-ion batteries in 1991.
WHY DOES THIS RESEARCH MATTER?
Many of the devices we use are powered and made possible by lithium-ion batteries. They are also commonly used in environmentally friendly electric cars. Improvements to these batteries continue to be made.
Nobel Prize in Chemistry press release: https://www.nobelprize.org/uploads/2019/10/press-chemistry-2019-2.pdf
© 2019 Compound Interest/Andy Brunning - compoundchem.com | @compoundchem
Shared under a CC Attribution-NonCommercial-NoDerivatives licence

Chapter 7 Fuel Cells
（燃料电池）

Find more information at HuaNeng, a public wechat account.

Part A Comprehensive Reading

Fuel Cells: History and Fundamentals

1 The History

The history of the fuel cell can be traced to experimental studies of the origins of electricity and its connection to chemical reactions, studies that were taking place at the end of the 18th and the beginning of the 19th centuries. One of the scientists involved in this endeavor was the British scientist Sir Humphry Davy. Davy studied the effect of an electric current on water and showed that the passage of a current could split water into hydrogen and oxygen, the process now known as the electrolysis of water. In 1802 he also devised a very simple electrochemical cell using oxygen that was able to deliver a small electric shock. This device had a carbon anode and nitric acid electrolyte but Davy did not publish his findings.

The history of fuel cells

While Davy's device is now recognized as a crude form of fuel cell, the first fuel cell that conforms to the modern description was a device that was developed independently by the German scientist Christian Friedrich Schonbein and the British scientist William Grove. Both men were exploring the idea that if water could be split into its constituent elements by an electrical current, then the process could be reversed and hydrogen and oxygen used to generate a current. Grove's letter on the subject was published in *The London and Edinburgh Philosophical Magazine* and *Journal of Science* in December 1838, Schonbein's in 1839 and in consequence Grove is now considered the father of the fuel cell.

Grove's fuel cell consisted of an electrolytic solution of sulfuric acid into which electrodes were placed. Hydrogen was delivered to one electrode, oxygen to the other in a cell that is similar in concept to the modern phosphoric acid fuel cell. Grove called the devices gas voltaic batteries although they were also called Grove cells. The actual name "fuel cell" was first used in 1889 by Charles Langer and Ludwig Mond who were trying to develop a practical cell from Grove's work using coal gas and oxygen. The cell never achieved practical success and research was largely abandoned in the 1900s after the development of the internal combustion engine.

The next milestone in fuel cell development came in 1930s when another British scientist, Francis Bacon, began to study the alkaline fuel cell. Bacon pursued his interest initially as an amateur inventor but later was able to organize research teams to develop the fuel cell. This culminated, in 1959, with the demonstration of a 5kW "Bacon Cell". In the same year, engineer Henry Ihrig, working for the Allis Chalmers Manufacturing Company, built a 15kW fuel cell stack of 1008 cells based on the alkaline fuel cell and used it to power a tractor. The company later built a golf cart, a submersible and a fork lift truck powered by fuel cells.

Bacon patented the design he had developed and in the early 1960s the Pratt and Whitney Aircraft Corporation licensed the Bacon Cell and began to develop the technology. The company was able to develop an alkaline fuel cell that attracted the attention of NASA and was subsequently used in the Apollo space program. The company's fuel cells were able to supply both power and drinking water to the astronauts on manned space flights and were later used on the space shuttle too. Pratt and Whitney was eventually taken over by United Technologies Corporation (UTC).

In 1955, at the same time as Bacon was developing his fuel cell, Thomas Grubb and Leonard Niedrach, two scientists working for general electric (GE) developed a different type of fuel cell that used a proton exchange membrane as its electrolyte. This too attracted the interest of NASA which was looking for a power source for its Gemini space program, the forerunner of Apollo. The GE proton exchange membrane fuel cell was eventually adopted for later Gemini flights before being replaced by the alkaline fuel cell for Apollo. The cell was also developed for marine use and was adopted in the 1980s by the UK Royal Navy for its submarine fleet. The proton exchange membrane fuel cell, which is extremely light, has since subsequently attracted the attention of vehicle manufacturers. It is now the main type of fuel cell being developed for automotive power.

The oil crises of the 1970s spurred research into a range of alternative technologies for power generation that included fuel cells as well as most of the new renewable technologies in use today. The fuel cell research from that period laid the foundations for the development of several new types of fuel cell including the phosphoric acid fuel cell, the solid oxide fuel cell, and the **molten** carbonate fuel cell. The last two are high-temperature fuel cells that operate under conditions that allow hydrogen and oxygen to react much more easily than at room temperature. To this list of fuel cell types was added the direct methanol fuel cell in the 1990s, a device that is particularly suited to mobile and portable applications. More recent additions include the proposal to use a perovskite crystal as the electrolyte in a fuel cell and the development of microbial fuel cells.

While development of fuel cell technology expanded during the 1980s, the commercialization of the technology was held back by cost. The fuel cells being used by NASA were too expensive for general use so cheaper alternatives were required if the technology was ever to find a market. It was not until the 1990s that the first commercial stationary fuel cell was launched, a phosphoric acid fuel cell system developed by International Fuel Cells, a subsidiary of UTC and Toshiba. There were also developments in Japan at around the same time based on the same fuel cell technology. However, fuel cells remained relatively expensive and it was not until the following decade that the costs fell sufficiently for a commercial market to begin to develop. Since then the market has expanded, slowly, for all the main types of fuel cell.

2 Fuel Cell Principle

The principle upon which the fuel cell operates belongs to a branch of chemistry called electro-chemistry. This explores how electricity can be derived from a chemical reaction. In nature certain materials will react with one another spontaneously if the conditions are correct. For example, a strong acid such as sulfuric acid will react vigorously with a variety of different metals if they are mixed with it. All that is necessary is to place a piece of the metal into the acid. Other reactions will proceed spontaneously but require a kick start. Typical of these is the reaction between hydrogen with oxygen. The two gases can be mixed at ambient temperatures without any reaction but once the temperature is raised beyond a certain point, with a spark for example, the reaction will proceed explosively.

Spontaneous reactions occur in nature because the products of the reaction are more stable than the starting materials. Reactions of this type are called exothermic reactions because they all release heat energy when they are allowed to proceed freely. This heat energy is equivalent to the difference in stability between reactants and products. In order to reverse the reaction, this amount of energy—or more—would have to be provided to the reaction products to push the reaction backwards.

Fuel cell principle-part I

All the chemical reactions that can be used to generate electrical energy in batteries and fuel cells are **spontaneous** reactions. When such a reaction is exploited electrochemically, the progress of the reaction is managed in such a way that some of the energy that would normally be released as heat emerges instead as electrical energy.

Fuel cell principle-part II

The reaction between hydrogen and oxygen is **exothermic**. This reaction can be expressed by a simple chemical formula:

$$2H_2 + O_2 = 2H_2O$$

The formula shows two hydrogen molecules and one oxygen molecule reacting to create two molecules of water. Although the formula looks simple, this is in fact a relatively complex process but it can be broken down into three simple partial reactions, each of which must take place for the reaction to run to **completion**. The first of these involves the hydrogen molecule, H_2, splitting into two hydrogen atoms, H, and the each of these releasing an electron to form a positively charged hydrogen ion, a proton:

$$H_2 = 2H^+ + 2e^-$$

A parallel second partial reaction involves the oxygen molecule, O_2, which splits into two oxygen atoms, O. Each of these absorbs two electrons released from two hydrogen atoms to produce a doubly negatively charged oxygen ion, O^{2-}:

$$O_2 + 4e^- = 2O^{2-}$$

The third and final part of the reaction involves the negatively charged oxygen ion attracting two positively charged hydrogen atoms and the three ions **coalescing** to form a water molecule, H_2O:

$$O^{2-} + 2H^+ = H_2O$$

Then the reaction is complete.

This reaction will not take place spontaneously at ambient temperatures because a significant amount of energy is required to rupture the bonds holding the oxygen atoms together in the oxygen molecule and the hydrogen atoms together in the hydrogen molecule. At high temperatures, these molecules will begin to dissociate spontaneously. Once dissociation begins and free oxygen and hydrogen atoms are available, these will react together very rapidly, releasing in the process much more energy that was needed to rupture the molecular H_2 and O_2 bonds. The high temperature in a spark is sufficient to initiate the reaction. However in a low-temperature fuel cell, the rupture of these bonds and the dissociation of the molecular hydrogen and oxygen must be assisted. This assistance is provided by a catalyst.

When hydrogen burns in air (oxygen), the various steps of the reaction occur in the same place at the same time, not in the stepwise fashion shown above. However in a fuel cell, the hydrogen and oxygen are not allowed to mix. Instead the reacting gases are introduced separately, hydrogen to one electrode of the cell and oxygen to the other. The two electrodes are separated by the electrolyte, as shown in Figure 1.

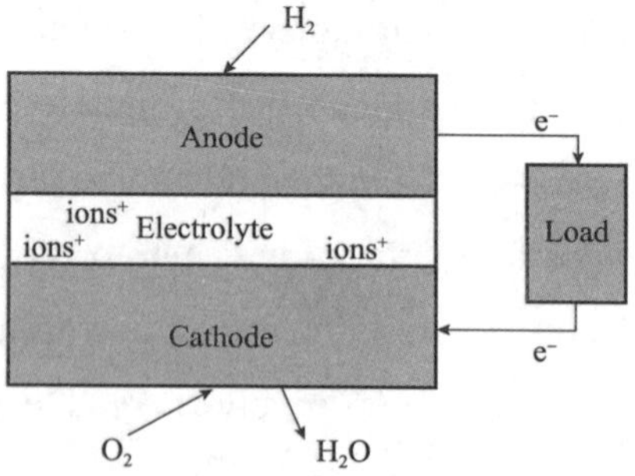

Figure 1 The fuel cell principle

The electrolyte is the key element in any electrochemical cell because it acts like a filter to both stop the cell reactants mixing directly with one another and to control how the charged ions created during the partial cell reactions are allowed to react with one another. The electrolyte in a fuel cell is **impermeable** to the gases, hydrogen, and oxygen. It will not conduct electricity in the form of electrons either, nor (provided it is an acidic electrolyte) will it conduct the negatively charged oxygen ions or the more common form in water, the hydroxyl ion (OH^-). What it will do is conduct positively charged hydrogen ions.

So, at the hydrogen electrode (which is called the anode) the hydrogen molecules first adhere to the electrode material as molecules and then separate into atoms, each subsequently releasing an electron to form a positively charged ion as shown in the equation above. In this ionic form the hydrogen can cross the electrolyte boundary and reach the oxygen at the second electrode. The electrons, however, are left behind at the anode.

At the second electrode (called the cathode), oxygen atoms will also adhere to the electrode surface as molecules, then dissociate, each leaving two oxygen atoms. However, these require a supply of electrons if they are to form oxygen ions. Only in this form can they coalesce with the hydrogen ions traveling through the electrolyte from the anode to create water molecules. The electrons must come from the anode, but these electrons cannot pass through the electrolyte.

If an electrical connection is made between the two electrodes, the electrons will pass through the connecting wire to the oxygen atoms where they will create oxygen ions, allowing the reaction to run to completion. When a small electric light bulb is inserted into this circuit, it will glow, proving that an electric current is indeed flowing between one electrode and the other.

All electrochemical cells operate in this way. While part of the reaction can take place within the confines of the cell, it can only be completed if electrons are allowed to travel from one electrode to the other through an external circuit. If no connection is made the exothermic nature of the reaction means that a charge **builds up** at each electrode, creating an electrical potential, the cell voltage, which will drive electrons from one electrode to the other when a connection is made.

Since the electrolyte is such an important part of an electrochemical cell, fuel cells are generally identified by the type of electrolyte they employ. Phosphoric acid fuel cells (PAFCs) use phosphoric acid, which being acidic is a proton (positively charged hydrogen atom) conductor. Proton exchange membrane cells also rely on an acidic membrane to allow hydrogen ions to pass. The alkaline fuel cell (AFC) is a hydroxide ion (OH^-) conductor; the hydroxide ion is the complement in water of the hydrogen ion. Meanwhile the high temperature solid oxide fuel cell (SOFC) uses a solid that is a conductor of oxygen ions. Other fuel cell reactions appear more complex even though the end result is the same. So, for example, the molten carbonate fuel cell (MCFC) electrolyte conducts carbonate ions.

3 Catalysts

The description of the operation of a fuel cell above is a simplification, because it omits one key feature of the reaction between hydrogen and oxygen. Although hydrogen atoms and oxygen atoms will react spontaneously to form water, both hydrogen and oxygen are found—at room temperature—in the molecular forms H_2 and O_2. These hydrogen and oxygen molecules must split into atoms before the reaction will proceed but they will not do so spontaneously because of the chemical bond holding each molecule together, even though these bonds are much weaker than the chemical bonds that will bind them together into a water molecule. The energy needed to cause the individual molecules to dissociated into atoms creates an energy barrier called the activation energy that must be overcome before the highly exothermic reaction between the atoms can take place.

Fuel cell catalysis

One method of splitting the molecules is to raise their temperature. They will start **dissociate** spontaneously at between 800°C and 1000°C. A flame or a spark will be hot enough to split sufficient of the molecules to start the reaction which then generates so much heat spontaneously that it keeps the reaction going until all the hydrogen or oxygen is used up. Some fuel cell designs also use high temperatures to encourage the gas molecules to dissociate but high temperatures bring their own design and materials problems.

The alternative is to use a catalyst. A catalyst is a component that is needed for a reaction to take place but that is not actually consumed during the reaction. It is usually used to accelerate a reaction that will otherwise take place very slowly or not at all. In the case of the fuel cell the best catalyst for low temperature acceleration of the reaction is the metal platinum. The platinum acts as a catalyst because both hydrogen and oxygen are attracted to the metal surface where they prefer to stick to it in dissociated form, as atoms, rather than as hydrogen or oxygen molecules. The metal can therefore provide the supply of atoms needed for the fuel cell reaction to take place. In its presence, the reaction can take place at below 100°C.

Fuel cell catalyst design-part I

Platinum, even though it is only required in small quantities in a fuel cell, is expensive and this helps to elevate the cost of the cells themselves. A key area of fuel cell research is therefore directed at finding cheaper alternatives. Platinum is also very sensitive to impurities in the gaseous reactants which can poison it, rendering it ineffective. Sulfur dioxide is a particular problem and so is carbon monoxide, both of which can find their way into hydrogen generated by the reforming of natural gas or other fuels. This is another reason why alternative catalysts, which are less sensitive to poisoning, are being sought.

Fuel cell catalyst design-part II

One area of research focusses on the exploitation of oxygen by natural species including animals and plants. It may not be immediately obvious but in fact both animals and plants exploit the same reaction as the fuel cell, using oxygen as a source of energy. However, animal and plant cells do not require platinum. Instead they use complex organic enzymes that contain metals such as iron and manganese. It may be possible in the future to use similar molecules to catalyze the reaction in fuel cells.

Fuel cell catalyst design-part III

4 Fuel Cell Types

There are six types of fuel cell that are commercially available today. These six are the AFC (alkaline fuel cell), the proton exchange membrane fuel cell (PEMFC, sometimes also called a polymer electrolyte membrane fuel cell), the PAFC, the MCFC, the SOFC, and the direct methanol fuel cell (DMFC). These cells are often identified simply by their acronyms. Some of the characteristics of these fuel cells are shown in Table 1.

Table 1. Characteristics of Main Fuel Cell Types

Fuel Cell	Operating temperature/℃	Typical size	Efficiency
AFC	<100	1~100kW	>60%
PEMFC	<120	1~100kW	60% with H_2 40% with reformed fuel
PAFC	150~200	5~400kW	40%
MCFC	600~70	300~3000kW	50%
SOFC	500~1000	1~1000kW	60%
DMFC	60~130	1~100kW	40%

Source: US Department of Energy, Fuel Cell Today.

The AFC is one of the most efficient fuel cells and operates at a relatively low temperature for a fuel cell. It was the first type of fuel cell to be fully developed and was used in the US space program. However, it has traditionally been considered expensive compared to other types. Recent advances are making it more competitive.

The PEMFC was also developed early in the modern history of fuel cells and it too was used in the US space program by NASA. The cell operates at a relatively low temperature and is very efficient when provided with pure hydrogen, less efficient when using reformed natural gas. The cell has received considerable attention and investment as a potential power source for vehicles due to its low weight.

The PAFC was the first fuel cell to be made available commercially, in the early 1990s. The technology was relatively expensive at that time and was expected to be superseded by cheaper alternatives but it made a strong comeback in the first decade of the 21st century and continues to compete with the other main types.

The MCFC is a high-temperature fuel cell that is able to reform natural gas internally. The fuel cell is more complex that the other main types and is only economically effective in larger sizes. However, it has been one of the most successful fuel cells for large stationary applications.

The SOFC is another high-temperature fuel cell. It is unique in having a solid state electrolyte with makes the cell potentially very robust. The fuel cell is suitable for both small and large installations.

The DMFC is the most recent addition to the stable of commercial fuel cells. Unlike the other fuel cells, it will "burn" methanol rather than hydrogen and this has made it attractive both for very small fuel cells for portable devices such as mobile phones and computers and as a potential power source for vehicles.

There are other types of fuel cell being developed but none has yet reached the stage where it is capable of being offered commercially. One of the most important branches is the development of microbial fuel cells that can exploit the reactions of microbes to produce electricity. Microbes use a variety of organic materials as fuel in order to generate energy. By diverting the normal reaction pathway in the same way as a conventional fuel cell diverts the natural way in which hydrogen and oxygen react, a microbial fuel cell can potentially use natural processes to produce electricity. These types of cell generally provide low power levels which is likely to limit their application to low power micro-electronic devices. Their advantage is that they can produce power from natural sources and so would not need providing with a source of hydrogen and oxygen.

Another area of interest involves the search for alternative solid state materials that can be used as electrolytes in fuel cells. Finding solids that will conduct ions but not electrons is the key and so far the range of materials has been limited. However, a class of crystalline materials called perovskites has recently been considered promising. There is still a long way to go before they might be used in commercial fuel cells but interest and investment is growing.

Selected from Chapters 1 and 2 of "*Fuel Cells*", Paul Breeze, Elsevier, 2017.

Notes

Molten: *adjective*. (1) (especially of materials with a high melting point, such as metal and glass) liquefied by heat. (2) denoting a cake or dessert that retains a thick, liquid center after being baked, typically served warm.

Spontaneous: *adjective*. performed or occurring as a result of a sudden inner impulse or inclination and without premeditation or external stimulus.

Exothermic: *adjective.* (of a reaction or process) accompanied by the release of heat. The opposite of endothermic.

Completion: *noun.* the action or process of finishing something.

Coalesce: *verb.* come together to form one mass or whole.

Impermeable: *adjective.* not allowing fluid to pass through.

Build up: assemble or accumulate something by putting parts or material together.

Dissociate: *verb.* (1) disconnect or separate (used especially in abstract contexts). (2) (with reference to a molecule) split into separate smaller atoms, ions, or molecules, especially reversibly.

Questions

(1) The main principle of fuel cell is that hydrogen react with oxygen to from water. Please describe how this process proceed?

(2) Please illustrate how the fuel cell work. What are the main components of a fuel cell?

(3) Why the electrolyte is the key elements in any electrochemical cell? What are the requirements of a good electrolyte?

(4) What are the tasks of electrodes? What kind of materials can be used as electrodes?

(5) Energy is needed to overcome the activation energy to let the reactions happen. What methods can be used to overcome the activation energy?

(6) Platinum-based catalysts are widely used in fuel cells. Please analyze the advantages and drawbacks of platinum-based catalysts. How to overcome these drawbacks?

(7) There are six types of fuel cells, which are commercially available today. What are they?

(8) The PEMFC was also developed early in the modern history of fuel cells and it too was used in the US space program by NASA. Please discuss what are the advantages and drawbacks of PEMFC.

Part B Reading for Innovation

Preparation of MEA

1 Introduction

PEFCs are considered a commercially viable product for both **stationary** and **portable** power, and largescale commercialization is predicted for the next few years, especially regarding hydrogen fuel cell vehicles. However, further efforts are required for progressive cost reduction, increased efficiency, the cell/stack increase in duration, and safety assurance. For this purpose, further research on materials, components, modeling, and new applicative studies and new test approaches are necessary for supplying industries with all the information needed both for developing products according to the market necessities, and for assuring good products that are compliant with quality standards. New concepts, materials, components, and design are first tested in laboratories before their application to large-scale systems. Consequently, the preparation of an appropriate testing set up, including a set of test protocols, is one of the most important activities in our laboratory's everyday practice.

This part is focused on preparation of membrane electrode assembly (MEA), the heart of the PEFC, for testing new materials, new components' architecture, or to supply information for modeling or for system management, and so forth. Hereafter, hydrogen-fueled PEMFCs are presented as the reference, taking into account that PEFC could also be fueled with alcohols and other fuels. Many different approaches and techniques can be used for the preparation of MEA in the laboratory according to the desired final characteristics, and based on available machinery. Indeed, some production methods that are very useful at the laboratory scale are not applicable at the industrial level, because they are not scalable, or they are too expensive in terms of time or costs. At the same time, some mass production approaches cannot be easily reproduced at the laboratory scale because they require costly and/or huge machines. For example, the sedimentation method is simple and relatively fast at the laboratory scale, and can be considered for small lots, but it is not advantageous for large-scale applications.

This part introduces general approaches and gives recommendations about MEA preparation without looking at **scalability**. The discussion will be centered on guidelines for the definition of a laboratory procedure. Attention is paid to ink deposition techniques, giving an overview of the most used and available, such as airbrushing and screen printing. Recipes and notes about the drawbacks, the industrial production processes, and future evolutions (ink-jet printing and 3D printing) are reported.

In this part, self-prepared or commercial available materials like, Nafion and Gore-Select membranes are applied. Additional information is provided on sputtering and chemical vapor **deposition** techniques related to onsite preparation of catalysts. The discussion includes the description of the MEA, the single-cell fixture for MEA testing and the MEA component preparation.

2 MEA and Test Holders

According to the IEC-62282-1 Terminology, the membrane electrodes assembly is a component of a fuel cell, usually PEFC/DMFC, consisting of an electrolyte membrane with gas diffusion electrodes (GDEs) on either side.

Figure 1 shows a schematic of an MEA section. The anode's gas diffusion layer (GDL) is composed of one or two sublayers, depending on the architecture (see Section 6.5). In the catalyst layer hydrogen is splitted into protons and electrons, the membrane allows the transport of protons to the cathode, in the cathode catalyst layer water formation occurs, and the cathode GDL is again built of one or two sublayers. Anode and cathode diffusion layers are composed of two sheets, these are usually called the support layer (SL), or macroporous layer and microporous layer (MPL), due to their different functions and porosities.

In operation, hydrogen and oxygen (or air) are supplied to anodes and cathodes, respectively. The produced water is removed by the gas flow in the anode and cathode channels. The produced electric current is collected and transported to the electric load (lamp, machine, electronic device, battery) by electric conductors.

For testing, an appropriate MEA test-holder is necessary, usually a single MEA per time is characterized in a "single cell" testing unit. The MEA test-holder interacts with the MEA, and its characteristics influence the experimental results.

In Figure 1, a sketch of the section of this embodiment is reported, and three common configurations are illustrated. The MEA is closed between two **planar** half-shells constituted of dense electric conductor materials that have channels for reactant gas adduction and removing exhausts; a couple of gaskets, usually made of synthetic elastomers, are used to avoid gas **leakage**. The two half shells are clamped by tie bolts or tie belts or, rarely, by an additional clamping device based on a hydraulic piston or a piston screw.

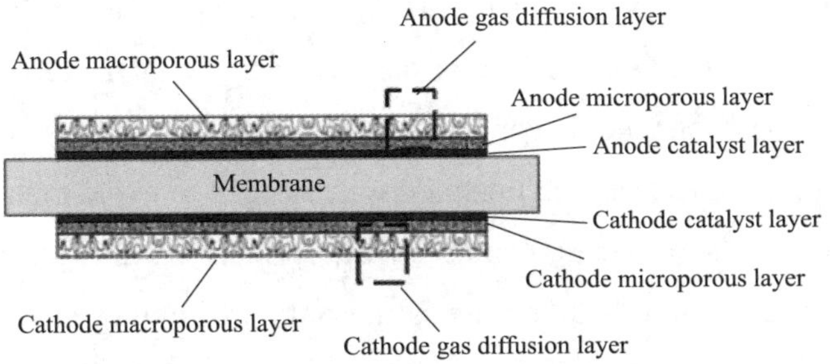

Figure 1 Schematic of MEA structure. Membrane and catalyst layers are sandwiched by two gas diffusion layers. Each diffusion layer is usually composed of two layers with different porosity called microporous and macroporous, or support layers, respectively

The two half-shells of the MEA holder are usually made up of three main configurations. In the most used configuration, Figure 2, each half-shell is composed of two parts: a plate, usually graphite, with channels for gas distribution (flow field) hollowed on the face in contact with the GDL of the MEA, and a metallic compression plate also working as a current collector. Tie bolts with insulating washers to avoid short circuits are used for clamping the cell, and a couple of gaskets are used to avoid gas leakage from cell sides. This configuration allows fast and low-cost replacement of flow field plates; the main issue is related to the metallic compression plate under voltage that can allow an accidental short circuit due to contact with other conductors.

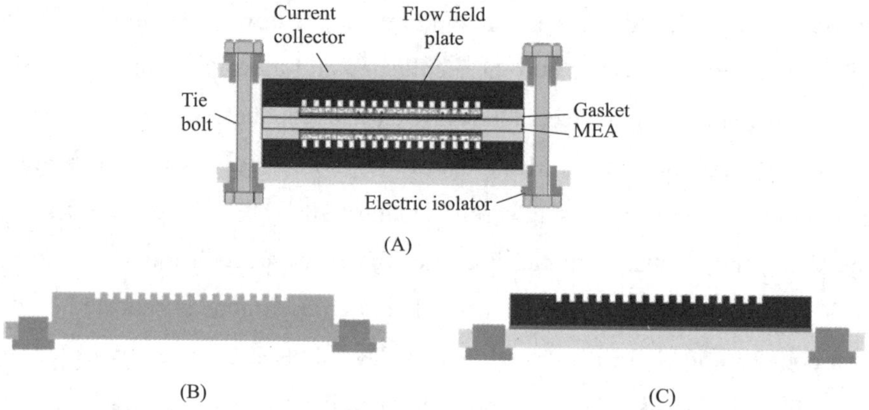

Figure 2 Schematic of single-cell enclosure. The most used configuration (A) foresees separate plates for the flow field and for compression/current collection. A single metal plate can be used (B) for reducing the number of pieces. In the configuration (C) two different plates are used for current collection and compression. In this last case, a compression plastic or other nonconductive materials plate could be used

The configuration reported in Figure 2(B) consists of a single metallic plate hosting both the flow field on the face in contact with the GDL, the holes for the tie bolt on its border, and the electric contactors for current drain. Due to the environment inside the cell, in this case, surface treatments are needed to avoid metal corrosion and surface passivation, also reducing contact resistance between this plate and the GDL. The very simple mounting/dismounting is the only advantage. The main issues are related to weight and accidental short circuits.

Another configuration, Figure 2(C) foresees the use of three plates: usually the one hosting the flow field is made of graphite a second one usually copper made is for current collection, and a third one is for clamping and it is made of stainless steel, aluminum, or plastic. Coming from the stack, this configuration allows a simple electric insulation of compression plates.

The MEA holder interacts with the MEA, mainly by the flow field . This means that changing the reference flow field also results in an MEA performance change. Moreover, the overtightening of clamping bolts, and the plates' deformations, negatively influence cell performance by damaging the MEA or increasing electric resistance.

After insertion of the MEA in the single-cell fixture, the test cell is connected to a test bench that has all the equipment for controlling gas fluxes, the operative conditions, and data acquisition and storage.

3 Electrolyte Membrane

The polymer electrolyte membrane (PEM) defines, to a large extent, the operative limits and performance of the cell, and, consequently, also the characteristics required of the other components for obtaining the best performance and endurance.

To reduce the cell resistance, and thereby increase the cell efficiency, it is useful to apply a very thin membrane. But this affects the mechanical resistance and the permeability to reactant gases, according to the properties of ions conducting polymer that is used for its preparation, and the preparation technique itself. Consequently, the selected membranes will be the best compromise between minimum **resistivity**, good mechanical characteristics, and low gas permeability.

The PEM is usually supplied ready for use as a foil or as a rolled tape, and for MEA preparation, it has to be cut with the correct shape and dimensions before use. The supplier gives information about the characteristics of the membrane, such as electrical conductibility, water uptake, swelling, operating temperature, glass transition temperature, and so forth. These data are useful both for MEA preparation, and experimental results analysis.

For the membrane, 3~15mm of extra border is added for sealing [Figure 3(A)] to the electrode area. The extra borders are in contact with the gaskets, which have a double function: sealing the borders of the catalyst and GDL layers, and compensating for their thicknesses, avoiding porous layers' damage due to mechanical compression stress. In common single cells, plain gaskets or O-ring-like sealings are used. The gaskets are introduced as a separate part, or are stamped onto the flow field plate, or onto the membrane.

The PEM's extra borders are not to be covered by electrodes and not to be exposed to mechanical stress due to gaskets. The common solution for protecting the PEM in this area is the addition of protective polymeric layers named pre-gaskets [Figure 3(B)]. The pre-gaskets are applied to the PEM's extra border by a thin glue layer, or by hot pressing, or rarely, by printing. The use of a glue, such as for adhesive polyester sheets, requires an accurate choice of this component to avoid catalyst poisoning, or spread of the glue out of the requested area. Hot pressing does not introduce additional chemicals, avoiding any possible catalyst pollution; in this case, attention must be paid to the maximum temperature and pressure that can be applied without damaging the membrane. For perfluorosulfonic membranes, a max of 150~160°C is recommended, while max pressure strongly depends on PEM thickness.

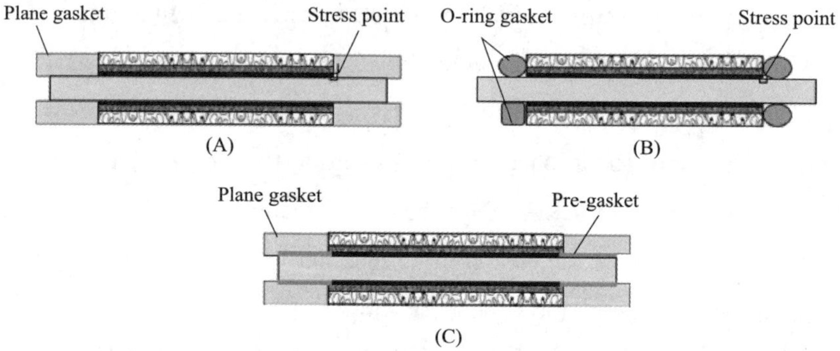

Figure 3 Schematic view of the section of an MEA and the related gaskets for a standard laboratory cell. With plane gasket (A) and O-ring (B), the electrode border introduces a stress on the membrane. The addition of pre-gaskets (C) reduces the membrane stress

For the membranes, often pretreatments according to PEMs' supplier specifications are required in order to remove possible traces of materials that could damage the cell performance:

(a) 30~60min in 3%~5% H_2O_2 aqueous solution at 80°C

(b) 1~2h in deionized water at 80~90°C

(c) Treatment in 0.5~1M H_2SO_4 aqueous solution at 80~90°C

(d) Washing in deionized hot water

(e) Storage in deionized water at room temperature

4 Catalyst Layer

The catalyst particles have to be in contact with both electronic and proton-conducting materials, and have to be accessible to the reactants. Moreover, the path for product water removal must be available in the CL. This means that three percolating networks must be present: one for electron conduction (catalyst materials), one for ion conduction

(ion conducting polymer), and a pore network for gas transporand water removal (Figure 4).

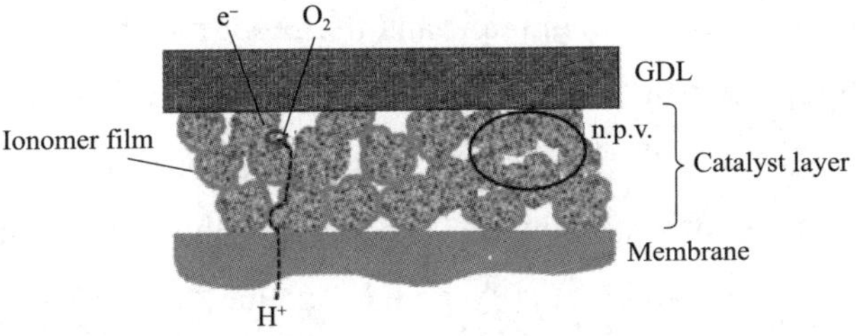

Figure 4 Scheme of the catalyst layer structure with reactant path at cathode

The ionomer creates ion-conducting films in contact with the catalyst grains this film allows the transport of protons from the membrane. Electrons coming from the bipolar plate pass through the GDL, migrate along the carbon grains supporting the metal catalyst reaching the reaction site, while oxygen reaches the reaction site by a percolating path of wet proofed pores, or pores that are not water filled. The reaction of water formation can occur only on catalyst grains where protons, oxygen, and electrons are simultaneously present. These sites are named the "triple phase boundary" (red circle, in Figure 4) and are realized where metal particles of the catalyst are in contact with the ionomer film, and there is access to oxygen. In nonpercolation volume (in Figure 4, "n.p.v."), no reactions can arise, because one of the percolating paths is not present; for example, in Figure 4, the ionomer film is not present. So too, at the anode, the hydrogen oxidation reaction is effective only at the triple phase boundary sites.

The "triple phase boundary" name comes from the electrochemical cells using a liquid electrolyte, because in this case, there is a liquid (the electrolyte), a solid (the catalyst), and a gas (the reactant) at the same point. In fuel cells with liquid electrolytes, the maintenance of the electrolyte film network without filling the pores is one of the major issues. With a solid electrolyte, a stable ion conduction network that is not filling pores is possible; however, due to penetration and distribution restrictions of the solid electrolyte, limitations in the utilization of the catalyst located in small pores are present. The issues related to the wettability of materials and structure are common to fuel cells with liquid electrolytes and to PEFC, because they influence the ion conduction and the availability to the pores of gases.

Finally, water produced during the reaction can be removed by a percolating path of pores, both in vapor and in liquid form, or can be absorbed by the solid electrolyte. In this last case, it can reach the membrane and the anode side.

The described CL structure is the result of extensive research on electrode development that started with the Teflon-bonded catalyst layer applied on the GDL (see the next section), a configuration also used in phosphoric acid fuel cells and other electrochemical devices, which finally arrives at the “catalyst coated membrane” (CCM) configuration. It has to be noted that the presence of percolating paths, like in the catalyst layer, are also necessary in other components of the MEA; yet only in the catalyst layer, three percolating paths are requested.

Therefore, in preparing the catalyst layer, the goal is the realization of a volume with three percolating paths without nonpercolating volumes, because these cause a reduction of the MEA performance and, in some cases, can trigger destructive processes. At the same time, it is needed to maximize the catalyst utilization and to optimize the transport of the reactant and the product removal.

For example, pores of 20~40nm diameter are challenging for Nafion ionomer penetration, and consequently, using supported catalysts (such as Pt on carbon), it is possible that the ionomer cannot reach the Pt nanoparticles located in small or deep pores of the support; therefore, these particles will be not useful for the reactions.

The selection of catalyst supported on mesoporous (5~50nm) carbons and low porous supports creating mesopores in the agglomeration process, like carbon nanotubes, increases the catalyst utilization.

However, PEFC catalysts usually consist of a powder of platinum (Pt)-based nanoparticles deposited on a carbon support, because a limited number of applications of unsupported metal particles are used.

After the catalyst selection, the catalyst layer preparation can start defining the preparation procedure. The first step is the choice of the route to follow. Starting from the catalyst powder, CL can be made onto the GDL or on the membrane, the first approach gives the so-called GDE and the second gives the CCM.

Although direct deposition/formation methods on membranes or GDLs have been proposed for catalyst layer preparation: for example, sputtering and physical-chemical deposition, these methods are special cases and not common practice in laboratory preparation, because they require dedicated machinery and rooms.

For preparing GDE and CCM, the starting point is the preparation of a catalyst ink, usually consisting in a suspension of catalyst particles in a solution containing the solid electrolyte ionomer that acts both as ion conductor and binder. Pore formers and other additives could be added to manage the CL architecture, to ameliorate the water management, or to increase the ink stability, just as examples.

The inner structure of the CL is determined mainly by on the ink composition, although the deposition method and related deposition parameters have an influence on the final structure.

The ionomer has a polymeric part (hydrophobic) and a polar part for ion conduction; consequently, the choice of the solvent for the ink preparation will change the ink properties, as shown in Figure 5.

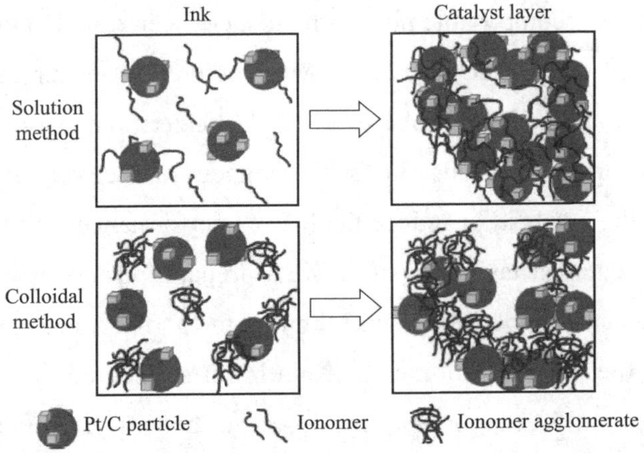

Figure 5 Scheme of the proposed ink structure (left) in solution (up) and colloidal (down) methods application, and related basic structures in the catalyst layer after preparation, respectively

Referring to Nafion, with its ionomer, a solution is obtained if the dielectric constant of the solvent is >10, while undergoing precipitation for a solvent dielectric constant of <3, and a colloidal solution is obtained for the dielectric constant in a range around 5.

The "solution method" for catalyst layer preparation uses solvents that allow a total dispersion of the ionomer to obtain an ionomer solution such as water, glycerol, and alcohols. While the "colloidal method" exploits solvents that allow the creation of ionomer aggregates, that is, the formation of a colloidal solution, such as esters and ethers. In the solution method, it is expected that the CL structure will be defined mainly by the catalyst particles, while in the colloidal method, it is the ionomer clusters that define the CL architecture.

In the experience of the author, and looking at the literature data, no significant differences in performance have been observed using the two different methods. The preference for one method or the other usually is related to the personal experience or to the selection of the deposition method, and rarely to the available or preferred base materials. Probably water management could be influenced by the two different structures, but literature data are not sufficient for assessing this aspect.

In both cases, to increase the CL **porosity**, a pore former can be added to the ink with the aim of obtaining greater, or more defined, porosity. Pore formers are materials that leaves void space for gas and water transport after their removal. Both water soluble and thermal decomposable pore formers have been tested, and can be removed both by a treatment of the electrode or the MEA before use, or directly inside the cell during the start-up and activation phase. Also, plasticizers and stabilizers are added to improve manufacturing and reduce ink sedimentation, respectively. All these additives must not poison the catalyst, or reduce the MEA's performance, and should be easy to remove after the catalyst layer, or entire MEA preparation; Otherwise, the additives could reduce reactant or product transport limiting the performance or the endurance.

Regarding the Nafion ionomer and platinum supported on carbon catalysts, 30%~35%, by weight, of the ionomer in respect to the catalyst is considered the best choice. This can be extended to other perfluorosulfonic polymers, while for other ionomers or other kind of catalysts, the best ratio could be different. This because the ionomer structure and the catalyst powder shape can be various.

At the end, we have a procedure for catalyst ink preparation:

(a) Weigh out the necessary 5% Nafion solution in a vial.

(b) Add the requested catalyst powder and mix by stirring, 1h and/or sonicate 5~30min.

(c) Under stirring, add isopropanol (or another alcohol or water) to reach the desired density and viscosity.

(d) Continue stirring until use.

By using 500mg of 5% Nafion solution and 50mg of 30% Pt/C catalyst powder, after drying, the catalyst layer will have a content of 33% of ionomer and 15mg of Pt (20%). Although this ink quantity could allow the preparation of $100cm^2$ of electrode with a platinum loading of $1.5mg/cm^2$, material that is loose due to the deposition technique reduces the surface that can be prepared, or the load of platinum.

The ink can be deposited according to a number of deposition methods on the GDL for GDE preparation, or on the membrane for obtaining CCM, or on a temporary support for a successive transfer of the desired host (decal method, used mainly for CCM).

Selected from Chapter 6 of "*Fuel Cells and Hydrogen*", Viktor Hacker and Shigenori Mitsushima, Elsevier, 2018.

Important Words and Phrases

Stationary: *adjective.* not moving or not intended to be moved.

Portable: *adjective.* able to be easily carried or moved, especially because being of a lighter and smaller version than usual. *noun.* a version of something, such as a small lightweight television or computer, that can be easily carried.

Scalability: *noun.* the capacity to be changed in size or scale.

Deposition: *noun.* (1) the action of deposing someone, especially a monarch. (2) the process of giving sworn evidence. (3) the action of depositing something.

Planar: *adjective.* relating to or in the form of a plane.

Leakage: *noun.* the accidental admission or escape of a fluid or gas through a hole or crack.

Resistivity: *noun.* a measure of the resisting power of a specified material to the flow of an electric current.

Porosity: *noun.* the quality or degree of having minute spaces or holes through which liquid or air may pass.

Part C Reading for Fun

Electrochemical Measurement Methods and Characterization on the Cell Level

1 Introduction

Electrochemistry is the discipline that combines the science of electrons in solids with that of ionic solutions. Depending on the applications, any material that will be used in electrochemistry needs fundamental measurements to insure its resistance, conductivity, reactivity, activity, stability, and durability in a considered medium. This chapter gives some electrochemical tools and techniques for basic measurements for studying the behavior of the electrode materials prepared. This chapter also provides the methodology for performing electrochemical measurements on three-electrode cells, as well as on fuel cell systems. From the cleaning step of the glassware to the measurement on the fuel cell, the different experimental methods are explained.

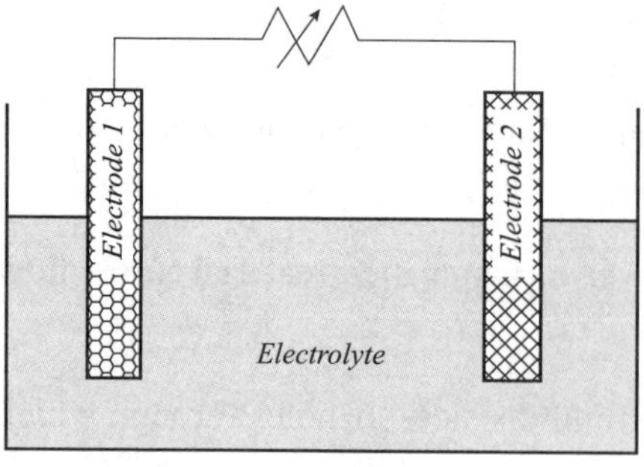

Figure 1 Schematic of an electrochemical system

2 Electrochemical Systems

A classic electrochemical system is composed of two electrodes immersed in the electrolyte, which is an ionic conductor (see Figure 1). This system can provide energy, or use energy for its operation.

Depending on the operation mode, a convention is required in order to identify the function of each electrode and the reactions that occur on them. Therefore, the electrode, where the oxidation reaction occurs, is called the anode, and the place where the reduction reaction occurs is called the cathode. In these two cases, the name of the electrode is based on its function. The name of the electrode is also based on its relative potential, in the same way that the fuel cells, primary batteries, electrolyzers, and secondary batteries are. Relative to the fuel cells and the primary batteries, the high potential electrode is the positive electrode (cathode, e.g., with ORR), and the low potential electrode is the negative electrode (anode, e.g., with HOR). In the case of electrolyzers and secondary batteries, the high potential electrode is the positive electrode (anode, e.g., with OER) and the low potential electrode is the negative electrode (cathode, e.g., HER). It is important for students in electrochemistry to understand these terms and to use them properly.

Electrodes are common elements of electrochemical measurements. The working electrode (WE) is the sample on which this chapter will focus. The counter electrode (CE) is the auxiliary electrode that allows the current flow. The reference electrode (RE) is the reference of the potential. It permits to control over the potential applied at the WE. The choice of the RE depends on the medium first, but also on the type of the electrochemical test to be performed. This choice is very important for a successful electrochemical measurement.

(1) From the Two-Electrode to the Three-Electrode System

The cell voltage can be expressed as follows in Figure 2.

where $U(I)$ is the cell voltage, $U°(I=0)$ is the theoretical cell voltage at $I=0$, η_{CE} is the change of potential across the metal/solution interface at the CE, and η_{WE}, the change of potential across the metal/solution interface at the WE. If the system is a battery or a fuel cell, the equation becomes:

$$U(I)=U° (I=0)-\eta_{CE}-IR-\eta_{WE}$$

and in the case of electrolysis (e.g., water electrolyzer), the equation is:

$$U(I)=U° (I=0) + \eta_{CE}-IR-\eta_{WE}$$

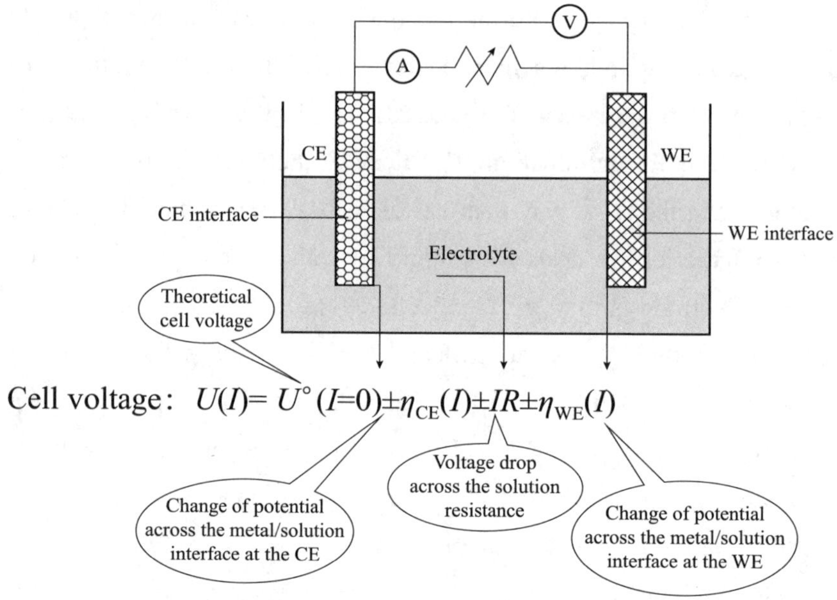

Figure 2 Illustration of the two-electrode cell and the different changes in potential

This configuration of the system makes it difficult, or impossible, to independently measure any change in potential at the WE, and that caused by the electrolyte. To achieve this, it is important to minimize the change of potential at the CE. For this purpose, the use of a CE with a surface area larger than that of the WE is recommended. In practice, the surface of the CE has to be at least five times that of the WE. Figure 3 shows the principle scheme of three-electrode cell for determining electrode potential.

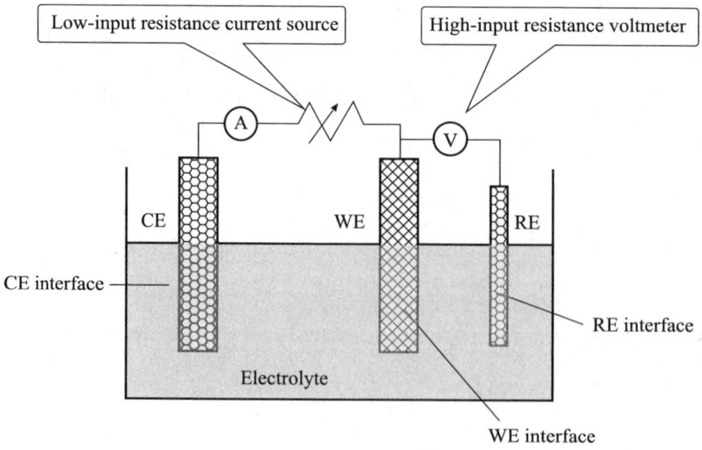

Figure 3 Principle scheme of three-electrode cell for determining electrode potential

Potential of the WE is compared with the reference electrode RE, which carries practically no current. The feature of the CE is such that the change in potential at this electrode is minimized. Thereby, the change in potential between the WE and the RE is equal to the change in potential of the WE. The potential of the WE is measured or controlled compared with the RE. The value of the WE potential is given by the following equation:

$$E(I)=E^{\circ}\ (I=0) \pm IR \pm \eta_{\mathrm{WE}}$$

Even if there is no current flow through the circuit used to determine the potential (WE vs. RE), in practice, the RE is often somewhere between the WE and the CE. Thereby, the potential measured includes some part of the potential drop between these two electrodes. Nevertheless, placing the RE very close to the WE can decrease considerably, but not eliminate, the potential drop caused by the solution resistance. For this purpose, the Luggin capillary is used in order to extend the ionic path from RE to the WE and to avoid any contamination of the electrolyte with anions such as Cl^-, SO_4^{2-}, and so forth, from the RE. Figure 4 shows the position of the capillary close to the WE. The best position of the Luggin capillary is its tip, which stays at a distance of 2d from the WE (where d is the diameter of the capillary tip).

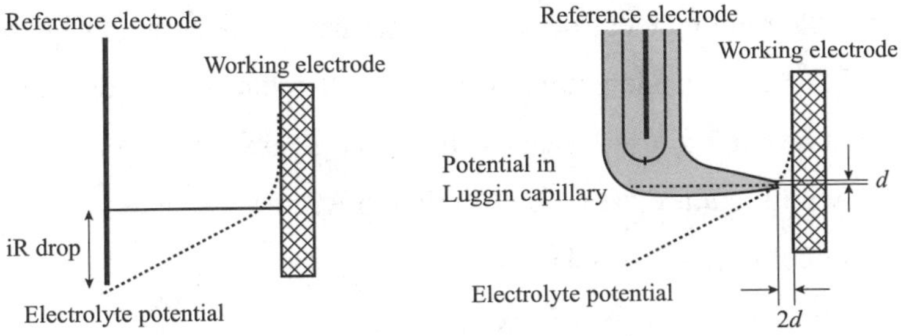

Figure 4 Ideal position of the Luggin capillary to the WE

(2) Examples of Three-Electrode Glass Cells

Figure 5 shows various three-electrode glass cells used in electrochemical laboratories. The choice of the cell depends on the experiment. The use of a Teflon or polymer cell to perform experiments in an alkaline medium is strongly recommended. An alkaline solution dissolves glass and leads to the adsorption of silicon component materials.

Figure 5 Various three-electrode glass cells

3 The Reference Electrode as A Cornerstone for Electrochemical Measurements

Given the major role of reference electrodes (REs) in electrochemical measurements, this section provides some thoughts on selecting the best one and then converting the obtained WE potential. In electrochemistry, the concept of “potential” comes from physics, and cannot be directly accessed. In the case of “potential,” only the potential difference between the electrode of interest (WE) and a reference, the so-called “reference electrode (RE)”, is measured. People running electrochemical measurements nearly always overlook the choice of an RE. The electrode potential range for the study depends on the nature of the RE you have decided to use. The next question is how to convert it to another RE for comparison with literature data. Some common REs are listed as follows:

Silver-silver chloride Ag/AgCl (SSCE):

$$AgCl_{(s)}+e^{-}\!=\!=\!=Ag_{(s)}+Cl^{-}_{(aq)}$$

Saturated calomel Hg/Hg_2Cl_2 (SCE):

$$Hg_2Cl_{2(s)}+2e^{-}\!=\!=\!=2Hg_{(liq)}+2Cl^{-}_{(aq)}$$

Mercury-mercury oxide Hg/HgO (MMO):

$$HgO_{(s)}+H_2O_{(liq)}+2e^{-}\!=\!=\!=Hg_{(liq)}+2OH^{-}_{(aq)}$$

Mercury-mercurous sulfate Hg/Hg_2SO_4 (MSE):

$$Hg_2SO_{4(s)}+2e^{-}\!=\!=\!=2Hg_{(liq)}+SO^{2-}_{4(aq)}$$

Silver-silver sulfate Ag/Ag_2SO_4 (SSSE):

$$Ag_2SO_{4(s)}+2e^{-}\!=\!=\!=2Ag_{(liq)}+SO^{2-}_{4(aq)}$$

Reversible hydrogen electrode $Pt/H_2/H^+$ (RHE) :

$$2H^+_{(aq)}+2e^- = H_{2(g)}$$

Overall, one of the most challenging issues is to find the most suitable RE, depending on the experimental conditions. An "ideal RE" should possess the following properties: (i) have a stable potential; (ii) meet the demands of a charge transfer imposed by the measuring instrument without changing its potential (be nonpolarizable); (iii) return to its fixed reference potential after accidental polarization; (iv) obey the Nernst equation for some species in a solution; and, (v) if it is an electrode of the second kind, the solid compound must be only sparingly soluble in the electrolyte. Examples of electrodes that possess all of these properties for use in aqueous solutions are the SCE and SSCE. The SSCE seems to be the scientists' "favorite RE" because of its simplicity, quietness, and robustness. It works at moderate temperatures up to 90°C, is filled with solutions of KCl, NaCl, and LiCl, and equilibrates fairly quickly when refilled. But if the electrochemical cell does not have a Haber-Luggin capillary tip, also known as a "salt bridge," that isolates the RE from the main body of the cell, the SSCE and SCE will not be the best choices. In some cases, not only can trace concentrations of ions (such as chloride) adsorb on the WE and dramatically change its electrochemistry, but they can alter the nature of the electrochemical reaction being studied. The job of the salt bridge is to allow only the ionic conductivity! If chloride ion issues arise, in which chloride must be rigorously excluded, the MSE can be considered. Furthermore, if the pH issue of solutions is taken into account, using either an SCE or SSCE electrode (often made out of plastic rather than glass) will not be a problem in acid media, but careful attention must be paid in alkaline media if the electrode has been in the base for a long time. After prolonged immersion, the pH of the internal filling solution could rise due to diffusion of OH^- into the RE leading to a potential drift because it is controlled by $Ag_2O_{(s)}+H_2O_{(liq)}+2e^- = 2Ag_{(s)}+2OH^-_{(aq)}$. To overcome that, MMO and RHE are recommended in alkaline media. RHE has the benefit of being filled with the same solution, and can be put directly into the test solution without adding trace impurities into the system, and obtaining a stable and well characterized reference potential. The measured potentials vs. the SSCE, SCE, MMO, and MSE reference electrodes can be converted to the RHE via the Nernst equations as follows:

$$E_{RHE}=E_{SSCE}+0.059pH+E^{\circ}_{(Ag/AgCl)}$$

$$E_{RHE}=E_{MMO}+0.059pH+E^{\circ}_{(Hg/HgCl)}$$

$$E_{RHE}=E_{SCE}+0.059pH+E^{\circ}_{(SCE)}$$

where E_{RHE} is the converted potential vs. RHE; E_{SSCE}, E_{MMO}, and E_{SCE}, are the experimental potentials measured against SSCE, MMO, and SCE reference electrodes, respectively; $E^{\circ}_{(Ag/AgCl)}$, $E^{\circ}_{(Hg/HgCl)}$ and $E^{\circ}_{(SCE)}$ are the standard potential of SSCE, MMO, and SCE at 25°C, respectively (values depend on the type of the filling solution).

Selected from Chapter 9 of "*Fuel Cells and Hydrogen*", Viktor Hacker and Shigenori Mitsushima, Elsevier, 2018.

Pt–Based Catalysts for the Oxygen Reduction Reaction

Pt is the most active single metal for oxygen reduction and is the current state-of-the-art catalyst for both cathode as well as anode in acid PEMFCs. It may reduce oxygen directly to water in a four-electron reduction at reasonable overpotential (see below). Initially, fuel cell cathodes were made from Pt black; however, attempts to lower the Pt content led to the introduction of carbon-supported Pt into PEMFC research, which are still the state-of-the-art in fuel cell applications. In these, Pt nanoparticles are supported onto conducting carbons (mainly carbon blacks) to enhance the Pt mass-specific surface area and Pt utilization, as there are relatively more Pt atoms at the surface compared to the bulk in smaller particles compared to larger ones. It has however been shown that Pt exhibits a particle-size effect in ORR with smaller particles having lower surface specific activity for ORR. The increase in active surface area is overcompensated by this loss in activity as soon as the particles get smaller than 4 nm, thus establishing an upper limit for Pt utilization, a significant issue due to cost reasons (see below). A particle-size effect in (electro)catalysis is usually attributed to the different activity of the various structural elements of the nanoparticle surface. For example, in a (111) surface, the atoms experience a higher coordination number by other surface and subsurface atoms than in a (100) surface. For a face-centered cubic lattice, for instance, this coordination number is 9 for the (111) and 7 for the (110) surfaces, respectively.

Atoms at corners or edges will have an even lower number of coordinated atoms. It is clear that these different levels of undersaturation of the surface atoms compared to the bulk ones will result in different catalytic properties. Thus, attempts have been reported in which preferentially structured nanoparticles have been synthesized, which exhibit a relatively large portion of the surfaces with the highest activities.

Application of such type of materials in electrocatalytic oxygen reduction showed a correlation of exposed facets in preferentially shaped particles with single-crystal results. However, although it becomes clear that the activity of Pt may be tuned in a certain way by advanced synthesis methods, doubts remain that the synthesized structures will remain stable under the conditions of aworking fuel cells, because, as will be discussed in more detail in the following, a fuel cell catalyst can be considered as a highly dynamic system and particle growth and restructuring may easily occur.

Various further attempts have been made to enhance the activity of Pt catalysts and at the same time enhance Pt utilization. The major strategy in this regard has for long time been the use of alloys of Pt with a second or a third metal. These developments date back to research on PAFCs and have been continued during PEMFC research. A broad variety of metal has been employed in binary and ternary Pt alloys, including Fe, Co, but also Ti, Ru, Mo, and Wo. While enhanced activity of Pt-based alloys compared to monometallic Pt has frequently been reported, the reasons for this enhanced activity have been a matter of debate for a long time. Both electronic as well as geometric effects have been considered to be responsible. Geometric effects consider a beneficial arrangement of the surface Pt atoms in terms of atomic distances and surface sites, which might differ in an alloy compared to the pure metal owing to different atomic radii, while electronic effects consider a change in the (d-)band structure of Pt due to the second metal.

An issue with such alloys is the possibility of leaching of the less noble metal, which might poison the proton-conducting membrane. The dynamic surface of such alloy catalysts and leaching may lead to structural features with properties different from those of a smooth, well-defined alloy. Markovic and coworkers investigated in detail the surfaces of alloys such as Pt_3Ni and Pt_3Co. They distinguished between two kinds of surface structures, namely, "skin" and "skeleton" structures. The authors were able to correlate oxygen reduction activity with the experimentally determined surface electronic structure, which is reflected in the d-band center. Activity is determined by interplay between adsorption energy of reactive intermediates on one hand and blocking of the surface by spectator species on the other hand. These relationships can be displayed in the form of so-called volcano plots, an example of which can be found in Figure 1.

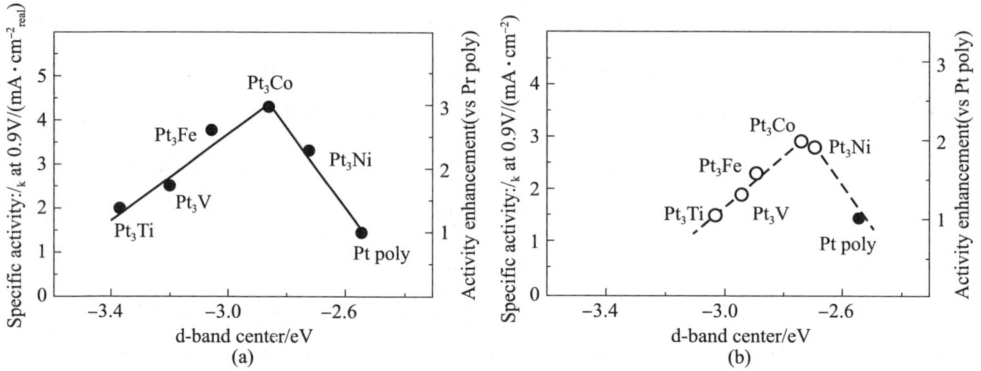

Figure 1 Relationships between the catalytic properties and electronic structure of Pt_3M alloys. (a,b) Relationships between experimentally measured specific activity for the ORR on Pt_3M surfaces in 0.1M $HClO_4$ at 333K versus the d–band center position for the Pt–skin (a) and Pt–skeleton (b) surfaces

Electrochemical dealloying starts from an alloy of Pt with one or more other metals. The PtCu and the PtCuCo systems are prominent examples and have particularly been shown to yield high ORR activity after dealloying. The electrochemical dealloying in principle is a simple electrochemical cycling between two potentials in a way that the non-noble metal of the outer layers of the material will dissolve into the electrolyte, thus a core-shell structure forms in which a monometallic shell (Pt) surrounds the alloy core. The nanostructure of the shell, composition of the material, and so on, all depend on a variety of factors, including particle size of the starting material, the alloy composition, and others. For example, it has been shown that multiple shell materials may form in larger particles (10~15 nm) with alternating zones of enrichment and depletion of the second metal (Co or Ni). Electrochemical dealloying of ternary PtNiCu or PtNiCo samples results in mass-specific activities up to five times that of a pure Pt catalyst. Recently, even six to seven times higher surface-specific activity and seven to eight times higher mass-specific activity compared to Pt catalysts has been reported for electrochemically dealloyed $PtNi_3$. The reasons for enhanced activity in such structures have been a matter of debate for some time.

Initially geometric effects, that is, the formation of structurally preferential atomic arrangements for ORR upon dealloying were postulated. However, meanwhile, it has become more and more accepted that strain-induced modification of the electronic structure/d-band center of the shell by lattice mismatch with the core is responsible for the enhancement in activity (see below). Figure 2 visualizes schematically the process of dealloying and the resulting core-shell structure.

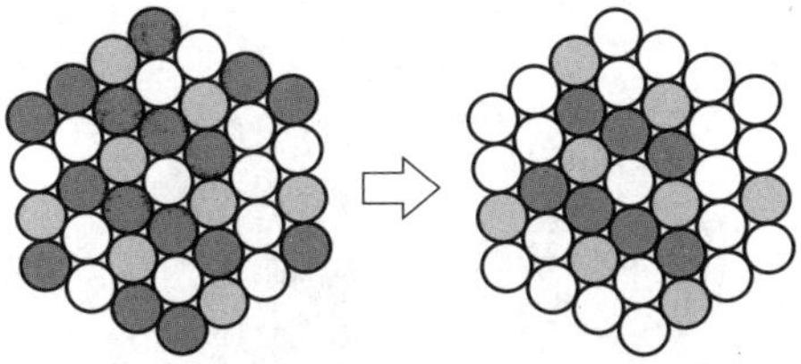

Figure 2 Schematic visualization of the formation of core–shell structures by electrochemical voltammetric dealloying of a Cu–rich alloy precursor (Pt, gray; Cu, red; and Co, blue)

The redox-exchange approach is mainly a two-step procedure [33, 34]. In the first step, a layer of, for example, Cu atoms is deposited onto the core metal in an UPD (underpotential deposition) step. The UPD leads to a well-defined monoatomic layer of the deposited metal around the core. The second step is the redox exchange of this monoatomic layer by a more noble metal, that is, the deposited Cu layer is oxidized by the more noble metal (e.g., Pt), which in turn is reduced into its metallic state.

Under appropriate conditions, this approach leads to the formation of a monolayer of Pt surrounding a core of another metal or an alloy. This approach would lead to a maximum utilization of the Pt as every Pt atom is at the surface of the catalyst particle, and thus would be extremely cost efficient if cheap metal cores are employed. Furthermore, electronic and geometric effects of the core may have an influence on the catalytic properties of the shell, which might thus be higher than that of pure Pt. Indeed, strain-induced shifts in the d-band center of Pt have been considered to be responsible and activities 20-fold that of standard Pt catalysts have been reported. This approach has been demonstrated for bulk single-crystal electrodes as well as for nanoparticles and a variety of core metals has been employed, including monometallic systems such as Pd or core-shell systems as cores.

Recently, other structures such as Pd nanowires have been employed as cores for the deposition of a Pt monolayer as well, and these structures have shown good activity in electrocatalytic oxygen reduction. Attempts have also been reported where several layers of Pt were deposited in order to fine-tune the activity. Indeed, it was shown in PtRu systems that two Pt layers, when deposited onto a Ru core, possess higher electrocatalytic activity for oxygen reduction than monolayers of Pt.

A sketch of the process of redox replacement as well as the resulting core-shell structure can be found in Figure 3.

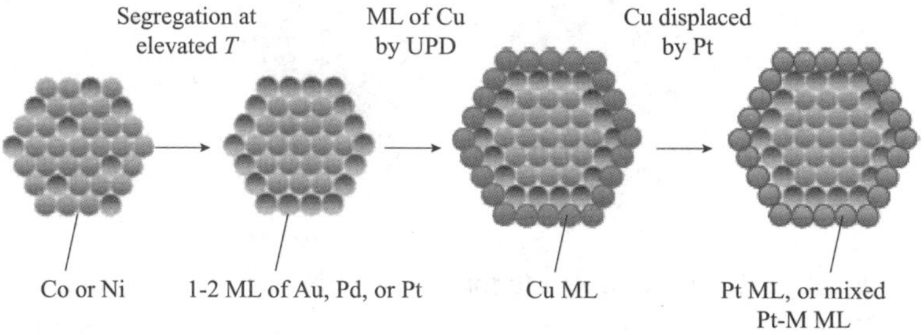

Figure 3 Schematic representation of the procedure for synthesizing Pt-monolayer catalysts on non-noble metal-noble metal core-shell nanoparticles

Besides the redox exchange of a monolayer of underpotentially deposited Cu, the direct redox exchange of the outer layers of a less noble core by a more noble metal has been also demonstrated and applied in the oxygen reduction. In any case, care has to be taken that a closed monolayer forms as recent reports have demonstrated that the redox-exchange approach may lead to open shells.

Selected from "*Non-Noble Metal Fuel Cell Catalysts*", Zhongwen Chen, Jean-Pol Dodelet and Jiujun Zhang, Wiley-VCH, 2014.

Chapter 8 Life Cycle Assessment（生命周期分析）

Part A Comprehensive Reading

Importance of Life Cycle Assessment of Renewable Energy Sources

1 Introduction

Progressive depletion of conventional fossil fuels with increasing energy consumption and greenhouse gas (GHG) emissions has led to a move toward renewable and sustainable energy sources. The production of sustainable energy based on renewable sources is a challenging task for replacing the fossil-based fuels to get cleaner environment and also to reduce the dependency on other countries and uncertainty of fuel price. A worrying statistic is that the global production of oil and gas is approaching its maximum and the world is now finding one new barrel of oil for every four it consumes. All these serious concerns related to energy security, environment, and sustainability have led to a move toward alternative, renewable, sustainable, efficient, and cost-effective energy sources with lesser emissions.

The life cycle assessment (LCA) of renewable energy sources is the key to observe their sustainability. There is a need to conduct LCA of renewable energy production system on the basis of their local conditions, as one energy source cannot be sustainable for all geographical locations, due to variations in resources availability, climate, environmental, economical and social conditions, policies, etc. Therefore, LCA can be used as a tool to assess the sustainability of various energy sources for different locations. LCA techniques allow detailed analysis of material and energy fluxes on regional and global scales. This includes indirect inputs to the production process and associated wastes and emissions, and the downstream fate of products in the future. LCA studies vary in their definition of the various criteria, such as, scope and goal, system boundaries, reference system, **allocation** method. LCA studies of renewable energy sources calculate the environmental impact and can relate the results against sustainability criteria. The present chapter is an effort to highlight the importance of LCA of renewable energy sources to get a more holistic perspective of their environmental sustainability.

2 Renewable Energy Sources

As shown in Figure 1.

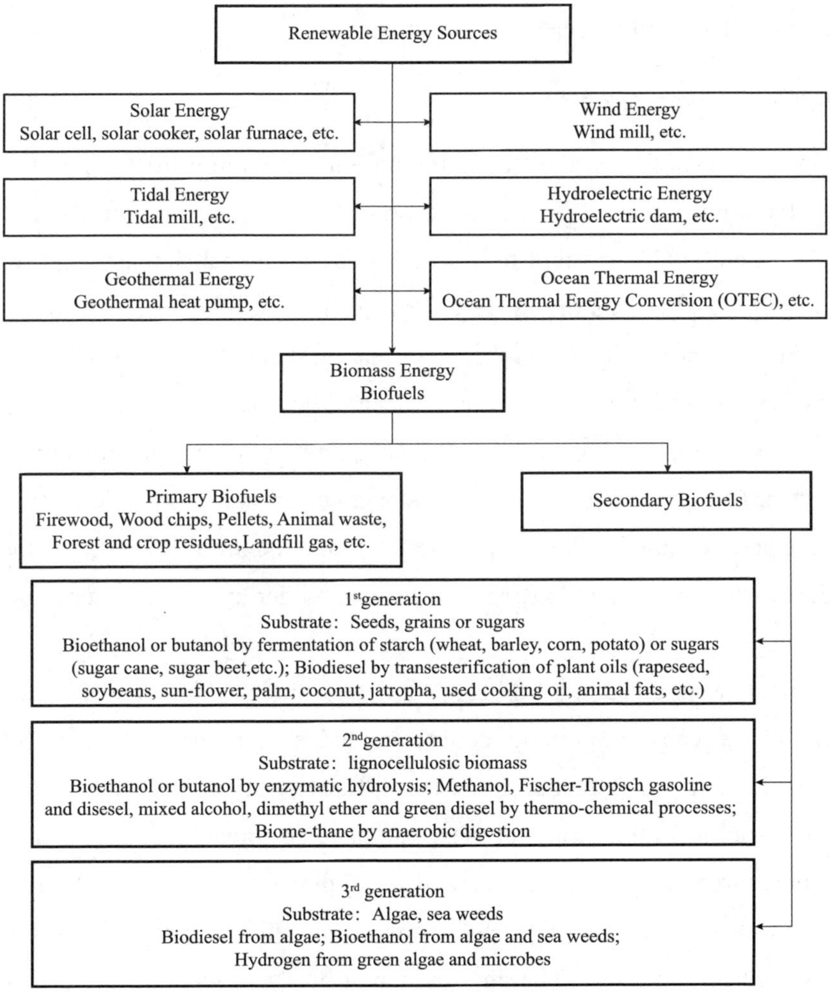

Figure 1 Biomass as a renewable energy source

The most common renewable energy sources are presented in the Figure 1. Each renewable energy source is performing differently; one could be best option for one location/purpose/season and could not perform with that efficiency at another location/purpose/season. The solar energy sources are best in remote or under developed areas having bright sunshine. Windmills are best suited near sea shore, as there winds are enough strong to get decent production of energy. Similarly, tidal, hydroelectric, geothermal, and ocean thermal energies have their importance. Among the renewable energy sources, biofuels are the most popular renewable energy source because of the availability of raw material (biomass), everywhere and round the year and also due to its suitability in transport vehicles and industries.

3 Life Cycle Assessment

ISO 14040 defined LCA as the "compilation and evaluation of the inputs, outputs and potential environmental impacts of a product system throughout its life cycle" (ISO 2006). Thus, LCA is a tool to assess the environmental impacts and resources used throughout a product's life cycle and consider all attributes or aspects of natural environment, human health, and resources and can be defined as a method for analyzing and assessing environmental impacts of a material, product, or service along its entire life cycle (ISO 2005). LCA analyzes the environmental burden of products at all stages in their life cycle (from the cradle to the grave) from the extraction of resources, through the production of materials, product parts and the product itself, and the use of the product to the management after it is discarded, either by reuse, by recycling, or by final disposal.

Various steps involved in the LCA methodology are listed in Table 1. The complete life cycle of the renewable energy sources includes each and every step from raw material production and extraction, processing, transportation, manufacturing, storage, distribution, and utilization. Each of these can have an impact (harmful or beneficial) of different environmental, economical, and social dimensions. It is therefore of crucial importance to assess the complete fuel chains from different perspectives in order to achieve sustainable biofuels. The environmental burden covers all types of impacts on the environment, including extraction of different types of resources, emission of

hazardous substances, and different types of land use. Attributional LCA describes the environmentally relevant physical flows to and from a life cycle and its subsystems, while consequential LCA describes how environmentally relevant flows will change in response to possible decisions. Marginal data are represented by the product, resource, supplier, or technology, which are the most sensitive to changes in demand, and economic value criteria are used to identify the marginal products.

Table 1. Overview of LCA methodological steps

Phase	Steps	Main result
Goal and scope definition	Procedure Goal definition Scope definition Function functional unit alternative and reference flows	Functional unit alternatives compared
Inventory analysis	Procedure Economy–environmental system boundary Flow diagram Format and data categories Data quality Data collection and relating data to unit processes Data validation Cutoff and data estimation Multifunctionality and allocation Calculation method	Inventory table, other indication (e.g., missing flows)
Impact assessment	Procedures Selection of impact categories Selection of characterization methods: category indicators, characterization models Classification Characterization Normalization Grouping Weighting	Environmental profile Normalized environmental profile Weighting profile
Interpretation	Procedure Consistency check Completeness check Contribution analysis Perturbation analysis Sensitivity and uncertainty analysis Conclusions and recommendations	Well-balanced conclusion and recommendations

Attributional LCA is limited to a single full life cycle from cradle to grave (see Figures 2 and 3 for this concept), and consequential LCA is not limited to one life cycle, but uses system enlargement to include the life cycles of the products affected by a change in the multifunctional processes will often be handled through allocation, physical flows in the central life cycle. In attributional LCA multifunctional processes will often be handled through allocation, while in consequential LCA, allocation will generally be avoided through the system expansion.

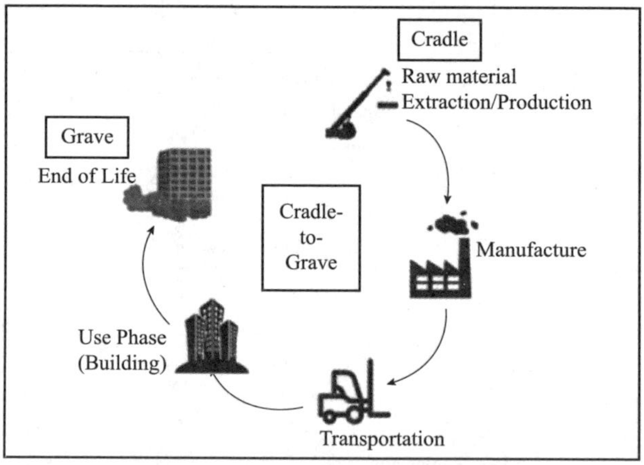

Figure 2　Cradle-to-grave life cycle assessment

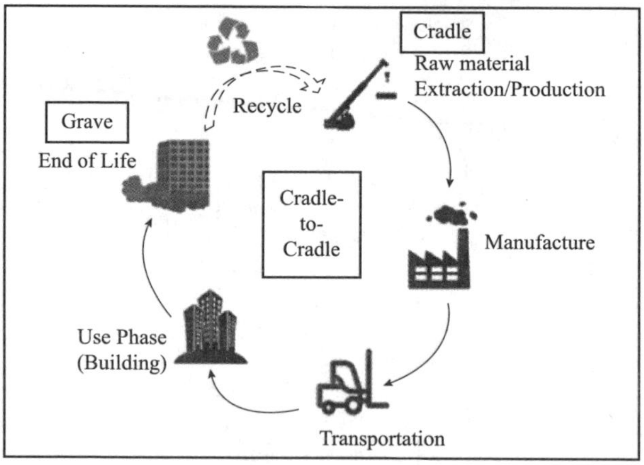

Figure 3　Cradle-to-cradle life cycle assessment

4　Sustainability of Renewable Energy Sources

The general principles of sustainable biofuel production are relatively easy to define (as shown in Figure 4). However, it is quite challenging to derive a sound framework that is able to characterize environmental, economical, and social impacts in an adequate way. World Commission on Environment and Development defined the term "sustainability" as "the development that meets the needs of the present without compromising the ability of future generations to meet their own needs" (UNCED 1992). The methodologies to address LCA and sustainability are advancing although the availability of practical data remains an issue. Sustainable development can be defined as the fulfillment through the optimal use of any available source within a production system.

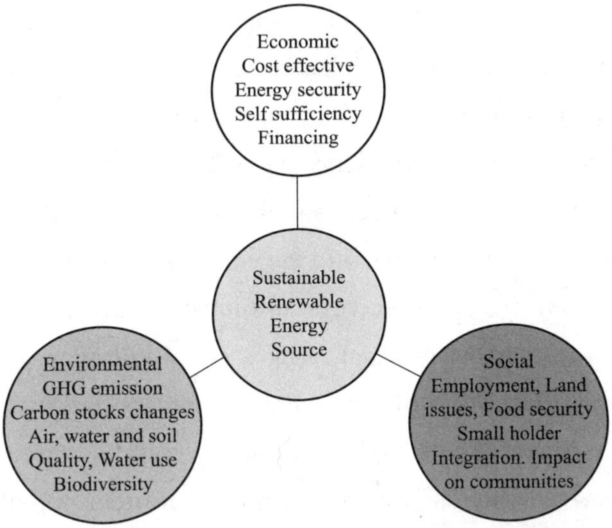

Figure 4 Economic, social, and environmental aspects of sustainable renewable energy sources

Energy conversion, utilization, and access underlie many of the great challenges associated with sustainability, environmental quality, security, and poverty. Sustainability assessment of products or technologies is normally seen as encompassing impacts in three dimensions, i.e., social, environmental, and economic. These three dimensions form the backbone of sustainability standards. To replace the fossil fuels with biofuels, there is a need to maximize the environmental and social value of biofuels that is also important for the future of biofuels industry and market potential depends on being cost competitive with fossil fuels (Figure 4). The environmental dimension comprises amongst others the greenhouse gas emissions, global ecological performance, conservation of energy resources, rational life cycle water use, effect on soil quality, conservation of biodiversity, use of chemicals, and the practice of slash and burn and the socioeconomic dimensions includes competition with food and feed, contribution to local well being, impact on communities and the quality of working conditions. These three interrelated goals must stay in balance for biofuels to remain sustainable.

Environmental impacts occur in all stages of the energy production system: the transformation of the land needed, production and application of chemicals and other input, cultivation of energy crops, production of the biofuel, transportation to the gauging station, and use in the vehicle. Pollutants are generated in many different steps of the production chain. The sustainability of renewable energy production depends on the net energy gain fixed in the output that depends on the production process parameters, such as the amount of energy-intensive inputs and the energy input for harvest, transport and running the processing facilities, emissions and their production cost. The most used indicators to measure the energy sustainability include life cycle energy balance, quantity of fossil energy substituted per hectare, co-product energy allocation, life cycle carbon balance, and changes in soil utilization.

The schematic illustration of the technical biomass potential and constraints to the sustainable biomass potentials is presented in the Figure 4. The technical potential of biomass is much lower than the theoretical potential due to cost involved in transport to collect them at production plant. The technical potential also has several social, economical, and environmental constraints, resulting only in a part of the technical potential that could be suitable for sustainable renewable energy production. Sustainability evaluation of biofuels is a multicriterial problem; Researchers have suggested the following main indicators for a sustainable energy production:

- To be carbon neutral.
- Not to affect the quality, quantity, and rational use of available natural resources.
- Not to affect the **biodiversity**.
- Not to have undesirable social consequences.
- To contribute to the societal economic development and equity.

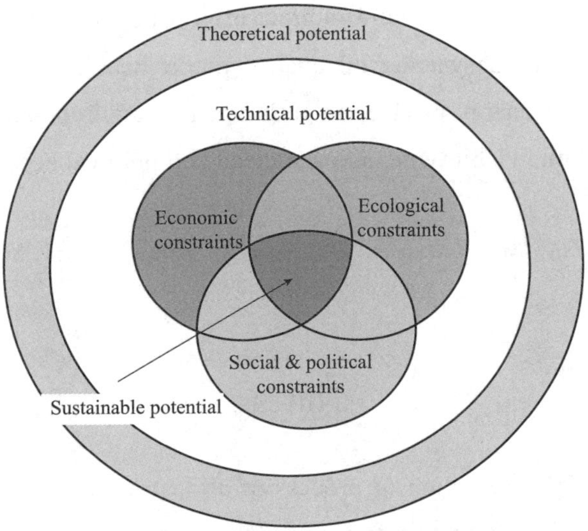

Figure 5　Schematic illustration of the technical potential and constraints to the sustainable biomass potentials

The major factors that will determine the impacts of renewable energy production system include their contribution to land use change, the feedstock/input used, technology adopted, scale of production, use of by-products (if any), **wholesale** trade and retail of energy product and by-product, and emissions after end use of produced energy. Studies have revealed that the interactions among various sustainability issues make the assessment of biofuel development difficult and complicated. The complexity during the whole renewable energy production chain generates significantly different results due to the differences in input data, **methodologies** applied, and local geographical conditions.The schematic illustration of the technical potential and constraints to the sustainable biomass potentials is shown in Figure 5.

In order to ensure net societal benefits of biofuels production, governments, researchers, and companies will need to work together to carry out comprehensive assessments, map suitable and unsuitable areas, and define and apply standards relevant to the different circumstances of each country. The length and complexity of the supply chains make the sustainability issue very challenging.

The main aim is to improve the performance of the strategies by enhancing positive effects, mitigating negative ones, and avoiding the transfer of negative impacts to future generations. The science of LCA is being stretched to its limits as policy makers consider direct and indirect effects of biofuels on global land and water resources, global ecosystems, air quality, public health, and social justice. The sustainable renewable energy production is directed by environmental impacts (direct and indirect), economic viability including societal and political acceptance.

Selected from "*Life Cycle Assessment of Renewable Energy Sources*", Singh, Pant and Olsen, Springer, 2013.

Notes

Allocation: *noun.* (1) the action or process of allocating or distributing something. (2) an amount or portion of a resource assigned to a particular recipient.

Hazardous: *adjective.* (1) involving or exposing one to risk (as of loss or harm). (2) depending on the effect of unpredictable and unanalyzable forces in determining events : depending on hazard or chance.

Biodiversity: *noun.* the variety of life in the world or in a particular habitat or ecosystem.

Wholesale: *noun.* a business may buy goods in large quantities directly from manufacturers, warehouse them, and resell them.

Methodologies: *noun.* (1) a body of methods, rules, and postulates employed by a discipline: a particular procedure or set of procedures. (2) the analysis of the principles or procedures of inquiry in a particular field.

Questions

(1) What is LCA?

(2) Please brief the general procedures the carrying out LCA and goals for each step.

(3) Could tell the difference between cradle-to-grave and cradle-to-cradle LCA?

(4) Please share your own understanding on economic, social, and environmental aspects of sustainable renewable energy sources. And please give some example for your illustration.

Part B Reading for Innovation

Life Cycle Framework and Techniques

1 Introduction to Developing Life Cycle Models

The specific purposes for a life cycle cost (LCC) perspective in **acquisition** management, product development, product upgrades, and so forth include:

(1)Estimating the total **ownership** costs (TOC) to the stakeholder

(2)Reducing/capturing TOC through use of LCC tradeoffs in the systems engineering/product development process

(3)Controlling costs through use of LCC contractual provisions in **procurements**

(4)Assisting in day-to-day procurement decisions

(5)Understanding TOC implications to determine whether to proceed to the next development phase

LCCs in general are viewed from a pre- and postproduction perspective, with typical main categories as shown in Figure 1. This partial list of LCC general categories can be used to develop more detailed costs. Every system is unique, and this is by no means an all-encompassing list.

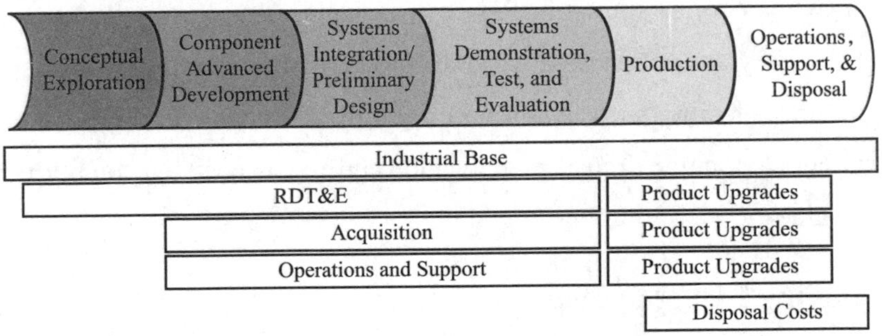

Figure 1 Some general LCC categories

2 Developing LCC Models

LCC analysis (LCCA) is an economic evaluation technique that determines the total cost of owning, operating, and disposing of a system over its life. When building a model for LCCA, you should consider two principal types of uncertainty:

- Uncertainty regarding the cost-generating activities.
- Uncertainty regarding the expected cost of these activities.

Both present unique challenges. In many respects, developing the categories is more challenging than estimating the costs. As W·Edwards Deming once said, “The greatest costs in business are unknown and unknowable.” This was the motive for developing and presenting a categorization methodology, with Figure 1 representing the top-level categories.

Once you develop the categories, the next step is to ascertain the costs and then develop an LCC model. A simple process for developing a LCC model is shown in Figure 2.

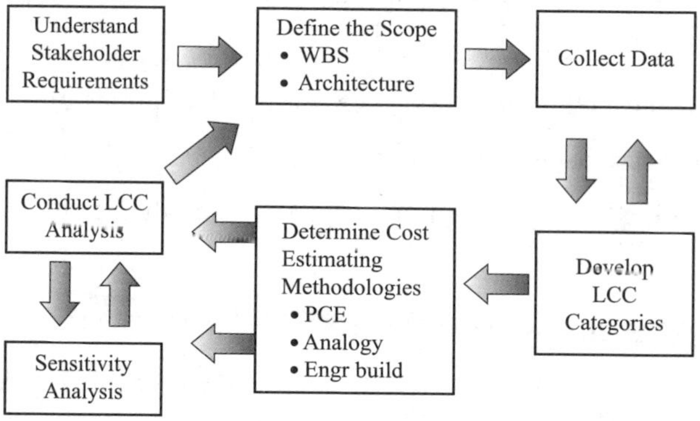

Figure 2 Process for developing an LCC model

The LCC categories shown in Figure 1 are applicable for the development of a large system. They are not applicable to all products. For example, consider an LCC analysis of a new facility such as a school. The main categories might include initial investment cost, operational cost, maintenance and repair cost, and finally residual or retirement value.

3 Developing LCC Models

Figure 1 shows some of the cost categories that should be included in developing an LCC model. The list is not all-**inclusive**. The figure is presented to demonstrate the complexity and time phasing of when to plan for and when the actual costs are incurred. The problem is further **exacerbated** because until the product goes into production we do not know the detailed hardware/software/interface components and thus have no way of developing life cycle categories and the appropriate costs. Thus, as we progress through the various phases in the product life cycle, we must update our LCC categories and costs. This section provides a brief discussion of each of the main categories shown in Figure 1.

(1) Industrial Base and Supplier/Vendor Relationships

Much has changed because of globalization. Many specialized, large-ticket items require an investment in the industrial base to support their value chain. This is especially true for defense systems such as tanks and submarines. Cultivating supplier/vendor relationships has become critical in an era when just-in-time inventory practices are the rule instead of the exception. New items may require primary contractors to invest in upgrading subcontractor capabilities to ensure quality and responsiveness. In our global environment, where offshore outsourcing and international partnerships are key, creating and sustaining the industrial base for major products while understanding and accounting for these costs throughout the product life cycle are important and significant. Table 1 summarizes some of these associated cost categories and elements.

Table 1. Industrial Base LCC Categories

Cost Category	Cost Element
Infrastructure investment	Workforce development and retention Physical infrastructure Minimum sustainment production
Other costs	Other costs

(2) Research, Development, Testing, and Evaluation

Research, development, testing, and evaluation (RDT&E) is the most commonly accepted term used to encompass development activities prior to production. From a development perspective (preproduction), these are the most difficult costs to ascertain because the product architecture has not been developed. Maturing technology for production is often impossible to plan and cost. Postproduction RDT&E costs are easier to ascertain because the makeup of the system is known. Table 2 lists some of the cost categories and elements used in developing an LCC model.

Table 2. RDT&E LCC Categories

Cost Category	Cost Element
Systems engineering	Systems engineering costs
Project management	Project management costs
Support and test program	Test and evaluation sets and expenses Training costs Data costs Demonstration costs Software and hardware development
Prime and subcontractor development	Prime and subcontractor infrastructure Licensing agreements
Product development costs	Hardware acquisition Hardware modification Software acquisition Software modification Software licensing Systems integration RDT&E
Contingency	Evolving requirements
Other costs	Other costs

(3) Acquisition

Once the product is designed, you can more accurately determine LCC because a detailed engineering bottom-up model is now possible. However, if you refer to Figure 3, you will see that 70% to 75% of the cost decisions are made during concept exploration. The problem is that you cannot develop realistic LCC because the subsystems and components of the system are not fully defined early in the life cycle. In other words, we cannot get a good handle on realistic LCC costs until late in the systems demonstration phase—once the costs are committed. At this point we can start to populate our LCC model. Table 3 is a partial list of cost categories and elements associated with acquisition or production.

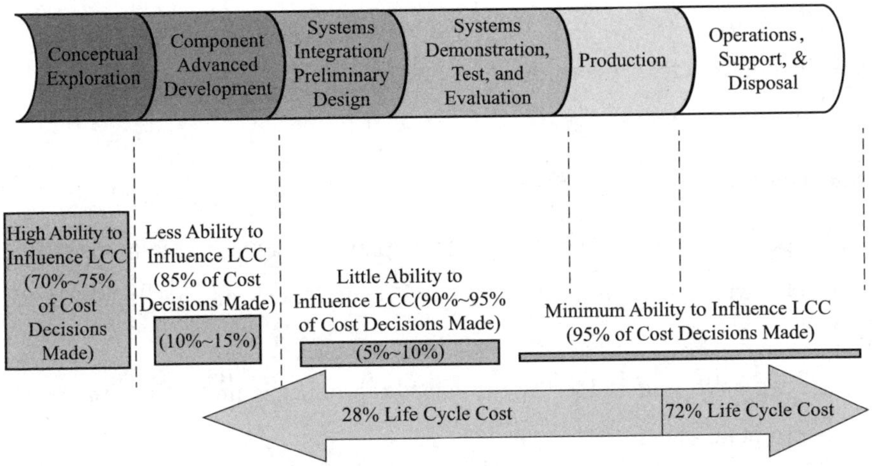

Figure 3 Costs incurred and committed during the systems life cycle acquisition process

Table 3. LCC Categories for Acquisition Expenses

Cost Category	Cost Element
Infrastructure	Physical plant Storage and spares Start-up cost Tooling
Product development	Initial item management Initial training Initial technical data
Inventory holding costs	Opportunity costs Finance costs Infrastructure
Production	Software acquisition Software modification Hardware acquisition Hardware modification Interface acquisition **Licensing** Warranty considerations Initial spares Transportation
Quality assurance	Test program sets and cost
Contingency	Evolving requirements
Other costs	Other costs

One item often overlooked is inventory-holding costs. These costs can range between 15% and 35% of the item value per year simply in lost opportunity costs, rent, pilferage, insurance, etc. Modern inventory management practices were created simply to minimize these costs.

(4) Operations and Support

Too often, operations and support (O&S) costs play a secondary role in the trade space studies used during concept exploration. We often focus on the development cost because:

• Once we have stakeholder buy-in and the product is under development, it can be hard to terminate the project.

• We simply do not know how to calculate LCC.

• The LCC is so overwhelming that many programs will never enter into production if the TOC influences the decision process.

Obviously, the operational life of a product drives the postproduction LCC. Table 4 presents rough orders of magnitude for purchase price as a function of TOC. As you can see, developing good postproduction costs for O&S is critical to capturing the TOC.

Table 4. Development Costs as a Function of TOC

Product	Purchase or Development Cost as a Percentage of TOC
Automobiles	30%~40%
Major defense systems	10%~20%
Commercial buildings	10%~15%

Note: These numbers do not include financing costs.

Table 5 lists LCC categories for the O&S phase. In building an LCC model, it is critical that you ascertain these costs, because of their relative contribution to TOC.

Table 5. LCC Categories for Operations and Support Expenses

Cost Category	Cost Element
Personnel costs	Labor maintenance costs Other support personnel Operational crews and management
Consumable goods	Energy Consumable components Some technology (e.g., personal computers)

(continued)

Cost Category	Cost Element
Maintenance cost	Overhaul support Spares consumption costs Training and management costs Spares replenishment costs Pilferage and damage of spares Infrastructure support Warranty and vendor maintenance
Inventory-holding costs	Opportunity costs Finance costs Infrastructure
Continuing system improvement	Software modifications Hardware modifications Integration/interfaces
Contractor support	Contract management
Sustainment support	Systems engineering and project management Product improvement programs/parts Documentation Value engineering Software maintenance/licensing Packaging and transportation Support equipment upgrades/replacement
Indirect support	Operational support personnel Operational personnel training
Infrastructure	Training facilities Support personnel facilities and costs
General training and education	Basic and initial skill training Education infrastructure
Acquisition	Acquisition
Other	Other costs

(5) Disposal or Retirement

Unless special conditions apply, planning for disposal costs is relatively straightforward. The problems arise in special cases such as asbestos, nuclear, funding retirement plans, and some drugs. For example, as of February 2005, Wyeth Pharmaceuticals had placed more than $16 billion in reserves for associated lawsuits for the diet drug Fen-Phen and had settled only 9000 out of 70000 lawsuits. Table 6 lists cost categories and elements for disposal or retirement of a product.

Table 6. LCC Categories for Disposal Expenses

Cost Category	Cost Element
Environmental	Cleanup Regulatory
Postproduction support	Personnel Spares Facilities
Retirement	Sell–off costs Storage costs Early retirements
Other	Documentation

Selected from Chapter 4 of "*Systems Life Cycle Costing: Economic Analysis, Estimation, and Management*", Farr, CRC Press, 2011.

Important Words and Phrases

Acquisition: *noun.* (1) an asset or object bought or obtained, typically by a library or museum. (2) the learning or developing of a skill, habit, or quality.

Ownership: *noun.* the act, state, or right of possessing something.

Procurement: *noun.* the action of obtaining or procuring something.

Uncertainty: *noun.* the state of being uncertain.

Inclusive: *adjective.* (1) including or covering all the services, facilities, or items normally expected or required. (2) not excluding any of the parties or groups involved in something. (3) aiming to provide equal access to opportunities and resources for people who might otherwise be excluded or marginalized, such as those having physical or intellectual disabilities or belonging to other minority groups.

Exacerbate: *verb.* make (a problem, bad situation, or negative feeling) worse.

Vendor: *noun.* a person or company offering something for sale, especially a trader in the street.

Licensing: *adjective.* relating to the granting of an official license.

Pilferage: *noun.* the action of stealing things of little value.

Part C Reading for Fun

Measuring Green Logistics Impacts

Introduction

Logistics decisions are metrics driven, which means that green logistics models and initiatives require having the right measurements of the various environmental impacts.

Pollution, noise, vibration, and waste measurements are technical in nature and can be estimated through specialized equipment. For example, the EPA or the European Union standards (see Table 1) are enforced by subjecting technologies to lab and road tests under standard conditions. Driving, weather, terrain, congestion, and operational conditions can dramatically change the actual environmental impact of freight operations. We were able to observe differences of more than 50% in total fuel consumption, NO_x and SO_x emissions under a variety of speeds and stop patterns. Environmental analysis of logistics operations needs to be very aware of the various assumptions underlying commonly used emission factors. This is particularly important when estimating GHG emissions.

Table 1. GHG emissions calculated for 1000 gal of fuel

Fuel	GHGs	Source	Emissions included	Results	Units
Diesel	CO_2	UK Defra	Pump-to-wheel	9998	kg CO_2
Diesel	CO_2	US EPA	Pump-to-wheel	10150	kg CO_2
Diesel	CO_2, CH_4, N_2O	GHG Protocol	Pump-to-wheel	10159	kg CO_2e
Biodiesel	CO_2	GHG Protocol	Pump-to-wheel	0	kg CO_2
Biodiesel	CO_2 (biomass)	GHG Protocol	Pump-to-wheel	9460	kg CO_2
Diesel	CO_2, CH_4, N_2O	GREET	Well-to-wheel	12933	kg CO_2e
Biodiesel	CO_2, CH_4, N_2O	GREET	Well-to-wheel	2964	kg CO_2e

(1) GHG Emission Measurement in Logistics

There are important variations in emission factors across sources (see Table 2). For instance, depending on the assumed (or observed) utilization factor of the conveyance and the granularity of data available (e.g., surveys or fuel records), the average fuel consumption per ton of cargo transported will vary significantly. Note that the EPA reference numbers are the same for trucks of various engine sizes, while the Defra numbers vary dramatically between large (over 33t) and smaller trucks (1.3~1.7t). This difference is due to the level of detail of data collected by these two agencies at the moment of publication of the GHG Protocol: the EPA was using sector-level aggregated data while Defra had access to vehicle-level surveys. Moreover, these numbers are regularly challenged and updated as more and better data becomes available across the logistics sector (Table 2). Another complexity, especially relevant to GHGs measurements, is what is and what is not included in the emission factors and calculations.

Table 2. CO_2 emissions for a 10000 short ton-mile shipment across multiple modes under GHG protocol

Emission factor	Value	Units	GHGs	Total/ ($kgCO_2$)
Air–long haul (US EPA)	1.527	kg CO_2/t–mile	CO_2	15270
Air–long haul (UK Defra)	0.346	kg CO_2/t–mile	CO_2	3460
Air–long haul (US EPA)	1.527	kg CO_2/t–mile	CO_2	15270
Air–long haul (UK Defra)	1.105	kg CO_2/t–mile	CO_2	11050
Watercraft–shipping–large container vessel (20000t deadweight) (US EPA)	0.048	kg CO_2/t–mile	CO_2	480
Watercraft–shipping–large container vessel (20000t deadweight) (UK Defra)	0.007	kg CO_2/t–mile	CO_2	70
Watercraft–shipping–small tanker (844t deadweight) (US EPA)	0.048	kg CO_2/t–mile	CO_2	480
Watercraft–shipping–small tanker (844t deadweight) (UK Defra)	0.019	kg CO_2/t–mile	CO_2	190
Road vehicle–HGV–articulated–engine size>33t (US EPA)	0.297	kg CO_2/t–mile	CO_2	2970
Road vehicle–HGV–articulated– engine size>33t (UK Defra)	0.049	kg CO_2/t–mile	CO_2	490
Road vehicle–light goods vehicle–petrol–engine size 1.305~1.74t (US EPA)	0.297	kg CO_2/t–mile	CO_2	2970
Road vehicle–light goods vehicle–petrol–engine size 1.305~1.74t (UK Defra)	0.462	kg CO_2/t–mile	CO_2	4620
Rail (US EPA)	0.025	kg CO_2/t–mile	CO_2	250
Rail (UK Defra)	0.016	kg CO_2/t–mile	CO_2	160

(2) GHG Standards and Scopes

There are three types of standards that cover GHG impact estimations. If the intent of the measurement is an absolute quantity for a whole company, this is known as Corporate Carbon Footprinting. The most widely known and adopted standard is the GHG Protocol followed by the ISO 14064. If the intent is to measure the GHG impact of an individual product, often from cradle to grave, it is known as Product Carbon Footprinting. The PAS-2050, ISO 14040, and the GHG Protocol Life Cycle Accounting and Reporting Standards are well-known references. Finally, and most recently, the EN 16258 (VTT 2012) standard for quantifying greenhouse gas emissions from freight focus specifically on carbon footprinting in the transportation sector.

One of the main differences, besides the goal itself, is the scope of the GHG calculations. Figure 1 shows all the life-cycle sources of GHG emissions of transportation services, from cradle to grave. These emissions are generated at different moments in time and points in space and allocated throughout the delivery of the service.

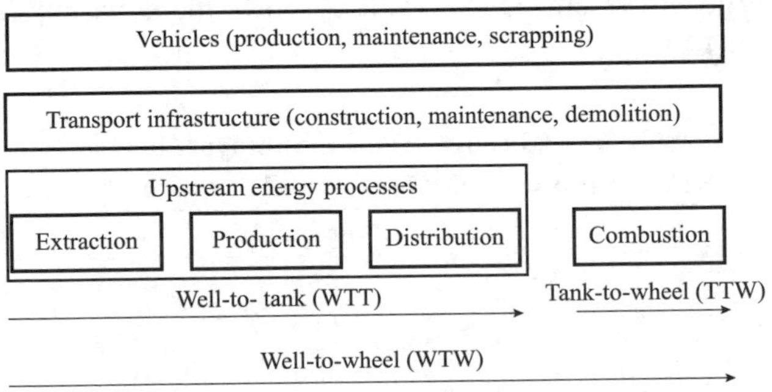

Figure 1 Life-cycle phases of transport

The first layer of emissions is all the GHG emissions generated during the manufacturing and maintenance of the trucks, airplanes, or boats. These emissions include the energy used in the assembly plants, the extraction of raw materials, and the operation of maintenance operations. Next are all the emissions related to the construction and maintenance of the transportation infrastructure, including roadways, ports, airports, and intermodal terminals. Some of this infrastructure is shared with other services (e.g., roads are shared with bicycles and personal vehicles, ports are shared with government and military operations) and needs to be properly allocated.

In order to be able to do a complete calculation of the impact of freight transportation, some share of the vehicle and infrastructure emissions should be added to each logistic operation. This is a fairly complex and uncertain calculation that requires very large amounts of data and assumptions. With the exception of lifecycle analysis or product carbon footprint calculations, most of these emissions are not included in GHG calculations in the logistics sector.

This leaves emissions from "well-to-wheel" (WTW): these are all the emissions related to the extraction, production, and distribution of fuels used during transportation operations, up to point when the fuel is placed in the vehicle—"well-to-tank" (WTT)—plus all emissions generated during the combustion of this fuel during transport operations, "tank-to-wheel" (TTW). The EN 16258 explicitly requires that all transport GHG calculations include WTW emissions. Other standards, including the GHG corporate protocol, only required TTW emissions, although WTW were encouraged. Similarly, the EPA Smart Way program initially only required TTW emissions and is now expanding to WTW emissions. Table 3 shows the emission factor differences between TTW and WTW by fuel.

Table 3. GHG emissions for various types of fuels (kg of CO_2 equivalents per L) in accordance with EN 16258

	TTW/(kg CO_2e/L)	WTW/(kg CO_2e/L)
Petrol	2.42	2.88
Ethanol	0.00	1.24
Petrol E5 (5vol% ethanol)	2.30	2.80
Petrol E10 (10vol% ethanol)	2.18	2.72
Diesel	2.67	3.24
Biodiesel	0.00	1.92
Diesel D5 (5vol% biofuel)	2.54	3.17
Diesel D7 (7vol% biofuel)	2.48	3.15
Compressed natural gas	2.68*	3.07*
Liquefied petroleum gas	1.70	1.90
Jet kerosene	2.54	3.10
Heavy fuel oil (HFO)	3.05	3.31
Marine diesel oil (MDO)	2.92	3.53
Marine gas oil (MGO)	2.88	3.49

Note:*per kg.

(3) GHG Allocation

The other source of complexity is the allocation of the emissions from a corporate or individual route level to an individual shipment. The GHG Protocol and the EPA program favor "simple" activity-based approaches: first compute all the emissions of a company or service using, for instance, fuel records. Then calculate the total distance (including empty miles) and weight moved during the transportation activities. Once WTW or TTW emission factors are applied to fuel consumption data, they can be divided by the total ton-mile of the logistics or transportation provider to estimate an emission factor per ton-mile. This approach ignores the fact that individual routes or services may have different efficiency levels, or that specific vehicles may have different performance metrics, in favor of a simple and consistent calculation that can be easily adopted by many organizations. The EPA SmartWay program does allow for tracking emissions by type of fleet (e.g., flat bed trucks vs. drayage trucks), but it still recommends an aggregate approach for computing emission factors.

The EN16258, on the other hand, favors a detailed approach. It recommends using the product of the weight of the consignment and the actual distance traveled—i.e., the transport capacity measured in tonne kilometers—as the allocation parameter. It aims at providing as accurate as possible assignment of emission factors to individual packages, taking into account logistics network configurations such as the relative location of the warehouse with respect to customers in a delivery route. Variations of this calculation are allowed in the standard, based on the quality of the information and type of service. An in-depth discussion of the EN16258 standard is beyond this chapter, but this requires consistently collecting data across every transportation route, including shipment information, sequence, and distance. This is a gargantuan task, especially since logistics operations often include multiple providers with various levels of sophistication. Nevertheless, for large organizations like UPS, DHL, FedEx, or Maersk, the EN16258 does provide a framework to develop information systems that can provide very accurate shipment level allocation of GHG emissions.

(4) GHG Metric Trade-Offs

As discussed on this section, GHG Metrics will vary due to assumptions, scope, or data availability. Standards can guide organizations' choices, but designing metrics that excel in all dimensions is not practically possible. Instead, firms must choose metrics that trade off between certain criteria. Two of the primary trade-offs are between integrative and useful metrics, and between robust and valid metrics (Figure 2).

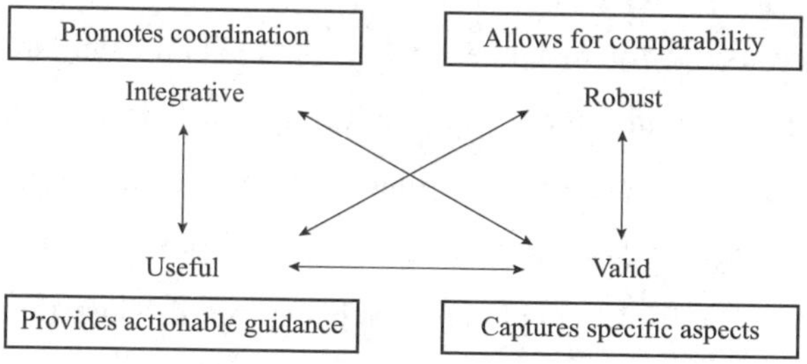

Figure 2　Trade-offs between criteria

The other trade-off is between a robust metric that allows for comparability and a valid metric that captures specific aspects. A valid metric, such as the ones favored by the EN16258, provides help with making a specific decision but is less suitable to external uses where it might be compared with similar metrics for other organizations, unlike the GHG Protocol or EPA approaches.

Selected from Chapter 7 of "*Sustainable Supply Chains: A Research-Based Textbook on Operations and Strategy*", Bouchery et al, Springer, 2017.

Moving from a Product-Based Economy to a Service-Based Economy for a More Sustainable Future

1　Leaving the "Product Comfort Zone"

The service sector has been an integral part of economic activity in both developed and developing economies. In 2013, services accounted for approximately 80% of GDP in the US. During the same period, in China the service sector continued to grow to 46.9% of GDP, a dynamic that may indicate the move to a new stage in China's economic growth.

Although such statistics are commonly used in texts describing the role of services in today's economy, another observation may help put a more interesting "face" on these numbers. In particular, for many years organizations that are (or were) typically recognized as manufacturing/product firms have been increasing the service component of their offerings. Examples include GM's successful creation of the OnStar subscription service or its expansion to financial services through the acquisition of AmeriCredit Corp. in 2010 to create what is known today as GM Financial. Similarly, SKF, the Swedish manufacturer of ball and roller bearings, now offers subscription-based diagnostic and predictive maintenance services. Dell and Hewlett-Packard created Dell Services and HP Enterprise Services, respectively, in order to provide IT and business services. These moves came in response to earlier moves by IBM, who has almost entirely transformed itself from a product-based company into a services company by selling its personal computers and servers business to Lenovo.

But why do manufacturers choose to go out of their "product comfort zone" to engage and invest in the creation of services? In the face of product commoditization and increasing competitive pressure, services can offer a stable and recurring revenue stream stemming from activities such as support, maintenance, and repair, which can extend well beyond the useful life of a product and generate lucrative profit margins. This can also increase the chances of cross-selling or alleviate customers' hesitation in upgrading to new products and equipment. Additionally, by offering after-sales support services, manufacturers can eliminate the need for third parties, who can erode brand perception, to perform these services. Furthermore, services are more difficult to reverse-engineer (i.e., to be imitated) and can facilitate longer and deeper relationships with customers due to longer contractual agreements. Manufacturers can also obtain a better idea about the customers' needs and the conditions under which the products operate and possibly customize their offerings accordingly. Finally, by offering services the manufacturers can acquire more accurate feedback about the performance of their products in the field, which can lead to product improvements and redesigns.

2 The Strategy of Servicizing

In the aforementioned examples, regardless of whether the manufacturers offer a support (e.g., maintenance) or an “add-on” (e.g., OnStar) service, the base of the transaction remains at the product level. This is because the service is contingent on the customers purchasing a product. However, in recent years we have observed a trend where manufacturers from various industries have started switching their focus from selling products to actually selling the solutions that their products offer. In most of the cases, manufacturers maintain the ownership of the products; therefore, the base of the transaction no longer seems to be the product per se, but rather the use of the product. This trend, also known as “servicizing,” indicates the transition from a product-based economy to a functional or solutions economy. In such an economy, instead of buying chemicals, for example, customers now have the option to buy chemical management services; instead of buying copiers, they can buy document reproduction services; and instead of buying cars, they can buy mobility services (Figure 1). Such options can be particularly attractive to customers because they free them from the administrative hassles and the operating risks (e.g., maintenance, repair) as well as the financial risks (e.g., depreciation of a product’s market value) associated with product ownership.

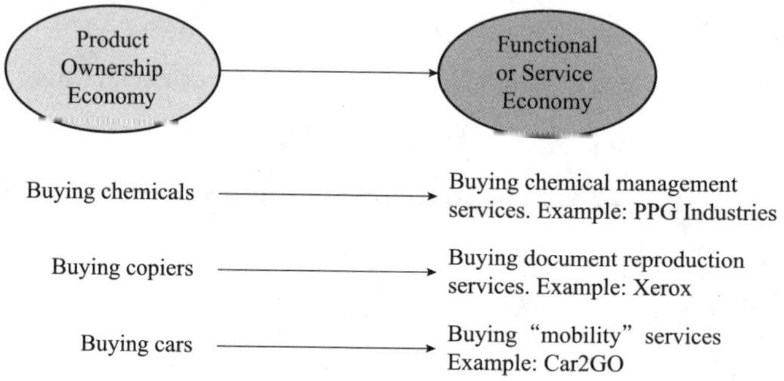

Figure 1 The transition from a product-ownership economy to a functional economy

The strategy of servicizing involves the transition from an existing business model to one where the focus is on the service or, better put, the solution provided to the end-customer. In fact, most business models can be thought of as comprising a combination of a product and a service component (Figure 2). A newly formed definition in the academic literature describes these models under the general term Product Service Systems (PSS) . The definition of PSS varies throughout the literature, but most authors